Jeffrey Lant
London
July, 1974

ALDERSHOT REVIEW

By the same author

SPLENDOUR AND SCANDAL
The Reign of Beau Nash

JOHN WALTERS

Aldershot Review

JARROLDS

JARROLDS PUBLISHERS (LONDON) LTD
178–202 Great Portland Street, London W1
AN IMPRINT OF THE HUTCHINSON GROUP
London Melbourne Sydney Auckland
Bombay Toronto Johannesburg New York

First published 1970

This book has been set in Bembo, printed in Great Britain
on Antique Wove paper by Anchor Press, and
bound by Wm. Brendon, both of Tiptree, Essex
ISBN 0 09 101890 0

To my memory of my brother,
Lieutenant Ernest Beauchamp Walters, B.A.,
aged 21, of the Gloucestershire Regiment,
killed at High Wood in the
Battle of the Somme,
30 July 1916

'Not the be-medalled Commander, beloved of the throne,
Riding cock-horse to parade when the bugles are blown,
But the lads who carried the koppie and cannot be known.'

JOHN MASEFIELD, *A Consecration*

Contents

Contents

Illustrations

Introduction

This is not a detailed chronological history of Aldershot. It is a review that covers the period between 1854 and 1935. We pass through those eventful years of the camp and the town, picking upon some of their strangest characters and incidents for inspection and study. And we follow some characters—both military and civilian—through their lives' parade, even before and after their sojourn at Aldershot.

The review starts with the activities of Albert the Prince Consort in the early 1850s. It ends with the last appearance of King George V at Aldershot in 1935, the year preceding his death. There passed with George the proud, pretentious but infinitely colourful and dramatic Aldershot. Then came new types of officers and men who preferred machines to horses and who, in the Second World War, fully justified the changes and reforms. The Camp today is a less important centre of the British army than of old. Nevertheless it remains a stimulating and exciting place, with an atmosphere of democracy and affluence among all ranks in incredible contrast to the facts related in my 1854–1935 review. And the townsfolk of Aldershot retain their traditional warmth and distinct individuality.

I thank a host of Aldershot citizens and past and present members of The Camp and district for many facts, figures and reminiscences. My special gratitude goes to:

Mr. E. G. Childs, managing editor of the *Aldershot News*; Major Onslow Dent, Public Relations Officer of the army's South-Eastern District; Major H. C. Ealand of the Devonshire and Dorset Regiment; Miss M. A. Killby, Superintendent of Miss Daniell's Soldiers' Home, Aldershot; Colonel G. H. H. Lee; Mr. W. N. Smyth, M.B.E., of Aldershot 'Old Contemptibles' Association; Mr. Ernest Stott, Public Relations Officer, Royal Aircraft Establishment, Farnborough; and Lieutenant-Colonel L. H. Yates, O.B.E., Superintendent, Prince Consort's Library, Aldershot.

I also gratefully acknowledge help from the Aldershot Public Library, with its excellent little civic and military museum assembled by the Borough Honorary Remembrancer Lieutenant-Colonel Howard N. Cole, O.B.E., T.D., author of the authoritative and comprehensive *The Story of Aldershot*; Farnham Public Library; National Army Museum of Royal Military Academy, Sandhurst; Imperial War Museum; British Museum Newspaper Library; and the French Institut du Royaume Uni.

The Detaille painting of the Prince of Wales and the Duke of Connaught is reproduced by gracious permission of Her Majesty the Queen and Cope's painting of the Duke of Cambridge by permission of the committee of the United Service Club.

The views expressed in this book are my own and are not necessarily shared by any of the persons mentioned above.

J.W.

Farnham, Surrey

I

The Camp is Born

THE quill pen of Albert the Prince Consort moved swiftly over a page of foolscap. He was starting to make some unsolicited comments upon the Defence Bill prepared by the administration of the Prime Minister, Lord Derby. The pen raced on until the prince's opinions filled twenty sheets of the long white paper. He drew attention to the perils that events in France had cast upon England and her fast growing empire. He demanded the establishment of a permanent army reserve force, arguing that reliance upon the present dribble of volunteers was not enough.

Queen Victoria shared the anxiety of her energetic husband regarding the ability of the new prime minister to overcome the many grave problems of this crisis year of 1852. Of course, the soldiers of the queen were always fighting a war somewhere, but this somewhere was a safe distance from the British Isles. Afghans, Sikhs and other warlike wallahs were too far away ever to vent their heathen wrath on British shores. However, just across the Channel an omnipresent threat had arisen in the vigorous and ambitious person of Louis Napoleon Bonaparte, nephew of the great Bonaparte vanquished at Waterloo. In December 1851 Louis Napoleon had forcibly dissolved the French Assembly and with much bloodshed had successfully achieved his *coup d'état*. The man's machinations had by December 1852 resulted in his being proclaimed emperor. This was three months after the death of the Duke of Wellington, the man who, it was believed, had disposed of Napoleonic tyranny for ever.

These events of 1852, so worrying to Victoria and Albert, had occurred during the administration of the fourteenth Lord Derby, a statesman of whom they disapproved. This prime minister was a racing man. His manners were often noisy and devoid of aristocratic dignity. Albert feared that Derby did not take his high office seriously, for he had often boasted, 'I'm too busy with pheasants to attend to politics.' The court might disapprove of Derby, but they could not despise him. This hearty shooting and racing extrovert had other gifts. He could converse with ease in Latin. He had translated the *Iliad* into blank verse with scholarly competence.

Prince Albert's views on the British army were widely respected. During Wellington's last years he had been so impressed by Albert's military gumption that he had begged him to become the commander-in-chief. The prince rejected this suggestion. He was aware of many unfriendly criticisms of his 'meddling' in the affairs of national defence. He preferred to work, as it were, behind the scenes, in close collaboration with his friend Colonel William Knollys. As titular colonel of the Scots Fusilier Guards, Albert placed himself under Colonel Knollys for practical training. Before nine in the morning on battalion field days in Hyde Park, Albert would arrive to be tutored by Knollys. He quickly became proficient at marching, drilling and in the use of arms.

Prince Albert's memorandum on the establishment of a permanent army reserve force was followed by a painstaking study of strategy. What if a French army landed on the Channel coast with the intention of taking over London? The prince's plan against such an eventuality was so modern that it was based upon the use of the railways. The London defence force would consist of 25,000 men. These would not all be herded in the capital but would have outposts near strategic railway stations. The first would be at Reading on the Great Western, the second on the South Western at Farnborough and the third on the South Eastern at Reigate. With such opportunities to jump on and off trains, the defenders of London would possess a mobility beyond even that of galloping horses.

Even if Albert's strategic concepts were sometimes questionable, his energy and enthusiasm regarding military matters never ceased to amaze. However, the prince's relentless and almost selfish drive had been a common topic of gossip among officers and politicians from the time of his marriage in 1840. His honeymoon with Victoria started at Windsor Castle, where seclusion was difficult because of the many guests then in residence. The first night of the honeymoon seemed unnaturally short, for early in the morning bride and bridegroom were seen fully dressed, walking briskly up and down the terrace. The spectacle caused Greville lugubriously to remark to Lady Palmerston that 'this is not the way to provide the country with a Prince of Wales'.

In succeeding years Albert maintained that the comforts of the bed must never interfere with his duty of contributing ideas for the safety and prosperity of the realm. When the clock struck seven a wardrobe maid would enter the apartments of queen and consort to pull aside the plush curtains and to raise the blinds. Albert would immediately leave his bed, cover his night-shirted body in a thick dressing gown and, if the air was damp and chilly, place upon his head a wig. Then— a strange-looking figure—he would hurry to his desk to wield his quill in an energetic session before dressing and eating breakfast. There were times when Albert became so inflamed by work that he gave the scratching pen precedence over his exalted spouse. Court flunkeys would disclose how her majesty, arriving at the door of Albert's study, would find it locked from the inside. 'The queen!' she would cry with imperious vigour. There would be no response. 'The queen!' she would repeat, with more passion; but again without a word from the husband on the other side of the door. Eventually queenly pride would dissolve until Victoria was crying softly and pitifully, 'Your wife, Albert.' Then Albert would lay aside his pen, rise and unlock the door, welcoming the queen with affectionate embrace.

To the satisfaction of Albert as military planner, the government of the worldly Lord Derby was defeated towards the end of December 1852 after only ten months of office. Now, in Lord Aberdeen, the queen and prince had the kind of prime minister they wanted. This coalition ministry was virtually Albert's own creation. Aberdeen was a serious, sober man with good manners, although at first encounter he repelled people by his grave and cold demeanour. Aberdeen was an aesthete, but without any aesthetic eccentricities that might shock the court. At the age of seventeen Aberdeen had sailed to Greece and wandered among its antiquities in ecstatic admiration. And Byron had written of him as 'the travelled thane'.

The man who succeeded Wellington as army commander-in-chief was also pleasing to the court. He was Lord Hardinge, a clergyman's son with an admirable military and political record. He fought at Corunna beside the fatally injured Sir John Moore. He lost his left hand at Quatre Bas. He was Secretary for War in the cabinet of Wellington. Yet all this later proved to be a prelude to colossal personal failure for Hardinge.

Prince Albert was soon realising that he had misjudged the situation in France. Her emperor, Napoleon III, was not a threat to England after all. On the contrary, he was doing his utmost to win England's trust and friendship. While agreeing that chances of war on English soil were becoming remote, the prince and Lord Hardinge continued to plan army reforms. The new anxiety in the nation's foreign relations was Russia. The hostile attitude of the Russians towards Turkey worried both the French and the British and brought them closer together. Both nations were opposed to any expansion by Russia into the domains of the sultan.

The prince was impressed by the vast training manoeuvres regularly undertaken by continental armies. He had another look at the manoeuvrability of British soldiers and decided they

were barrack-bound. They spent too much time in and around
the grim regimental buildings mostly situated in the heart of
bustling cities and towns. Conditions of actual warfare could
not, he argued, be simulated in barrack squares or even in the
available countryside in the vicinity of these. For frequently
movement of the soldiers interfered with farming and with the
grazing rights of the common lands. In addition, there were
garrisons who were quartered in ancient castles and forts,
archaic and confined. Under all such circumstances only rarely
could the various units of the army gather for manœuvres at
which actual warfare could be effectively imitated. Albert had
himself to obtain his British military training in Hyde Park, of
all places.

A training experiment made in 1852 was such a success that
now Albert and army leaders were dreaming of a more
ambitious plan. The experiment had been the formation of a
camp at Chobham, a few miles from what is now Aldershot.
There, under Lord Hardinge, troops fought sham battles on a
big scale over a large area of wild heath lands without the
disapprobation of or interference by civilians. A frequent
visitor to the scene was Prince Albert, usually accompanied by
his friend and instructor Knollys. The prince's blue eyes would
admiringly survey the 'battlefield' from his charger. These war
games were highly successful. The troops, it was officially
stated, were 'exhilarated and inspired' by the experience. But
actually they were made unhappy by perpetual rain, causing
Punch to comment: 'The men showed that they could not only
stand fire, but water too.'

Soon Hardinge and his staff were roving the heath lands of
Ash, Aldershot and Farnham, agreeing that as much of these as
possible should be possessed by the army. Prince Albert
discussed these recommendations with the queen. Again he
seized his quill pen to write long and detailed suggestions to his
malleable prime minister. Aldershot, the most famous military
centre in the world, had been conceived.

B

Aldershot was a village sprawling in desolate isolation three and a half miles from the ancient town of Farnham. The village had such a doubtful personality that some spelled it Aldershot and others Aldershott. There was a population of almost 900, but too scattered for such a total to be believed. A guide book of the decade listed but three residents as 'gentry', five farmers, two innkeepers, a blacksmith. Most of the houses were in the neighbourhood of Aldershot's only treasure—the twelfth-century parish church of St. Michael. The guide book recorded that in the Aldershot area there were 2,700 acres of heath and common, 731 acres of arable land, 19 acres of hop fields, 530 acres of meadow lands and pastures, 130 acres of woodland and 20 acres of buildings and gardens. There were few easy approaches to Aldershot. It seemed lost in wide expanses of gorse, heather and sand, in forbidding and desolate heath lands. The population of the area were infected by its wildness. Visitors passing along its bumpy highways found countryfolk of unkempt appearance and of unfriendly disposition.

Aldershot's two inns were the Red Lion and the Beehive. The Red Lion has survived in reconstructed form and can be seen today at the corner of Brighton Road and Ash Road in modern Aldershot. The Beehive also still flourishes. It is at the corner of Pond and High Streets. One day in August 1853 the Red Lion was thrown into a high state of curiosity by the arrival of an impressive-looking army general together with his personal staff. Accustomed ordinarily to hiring overnight beds to simple commercial travellers, George Foulkner, landlord of the Red Lion, and his wife were awed in recognising him as General Viscount Hardinge of Lahore, commander-in-chief of the army and fabled hero. They bustled about trying to make the shabby premises more worthy of such a personage while the surprised rustics stared fiercely over the tops of their beer mugs. However, Hardinge proved to be an easy and charming guest. He never lingered in the inn. Early in the

morning after a hearty breakfast he and his staff would be off
for the day on their horses.

One night in a parlour of the Red Lion Lord Hardinge wrote
a letter to Prince Albert recommending Aldershot as the site of
a permanent camp for the army. And in the following month
there came from Hardinge a detailed memorandum, with a
plan for the purchase for the army of some 10,000 acres.
Aldershot, he pointed out, was, from the point of view of
military strategy, ideally situated. It was the perfect spot for
assembling from all parts of the country a large military force.
Moreover, Aldershot was easily accessible from London, less
than forty miles away. Hardinge concluded the memorandum
with a plea for the earliest acquisition of the land. The government, pushed into a decision by Prince Albert's enthusiasm,
expressed support for the scheme. Within a few months
Hardinge was persuading owners of Aldershot lands to sell at
£12 an acre to the army. Soon £100,000 was allocated to buy
up land. This was later increased to £145,000. The army
became the proprietor of 8,000 acres, with more big acquisitions
to follow.

When Lord Hardinge and his staff busily surveyed the British
army's future domain, the Prince Consort was often away in
Scotland, materialising another dream of the queen and himself. In September 1853 he was at her side when she laid the
foundation stone of the new Balmoral with a libation of oil
and wine. But the prince's thoughts were miles away with the
army rather than at Balmoral. Through the medium of his
overworked quill pen he urged that the development of the
Aldershot camp should proceed with maximum speed. The
army, he wrote, must exercise there as soon as possible and on
the grandest scale. He foresaw the war which the nation entered
on the 27th of March in the following year. On that day
Prime Minister Lord Aberdeen 'unsheathed England's rusty
sword' against Russia in defence of the Sultan of Turkey. Then
when the war had come Albert seized the opportunity of

making Aldershot the great and permanent centre of army
training and manœuvres. He told the War Office that in view
of the public's support of this, the Crimean War, nobody
would dare oppose the development of a centre where the
troops could be properly and adequately trained. 'You can now
ask parliament for anything you want,' he added. 'Strike while
the iron is hot!'

The War Office struck. Soon there was row upon row of
tents, followed by 1,200 huts filling an area that had been
ragged heath land. Labourers were feverishly erecting barracks
and other constructions of brick and wood. Local residents
who had gloried in the heath lands' wild if desolate beauty
were almost frightened at what was happening. One such
eyewitness was Mrs. Ewing, writer of stories for children. This
is how she described the advent of The Camp:

'Take a highwaymen's heath. Destroy every vestige of life
with fire and axes, from the pine that has longest been a land-
mark, to the smallest beetle smothered in smoking moss. Burn
acres of purple and pink heather, and pare away the young
bracken that springs verdant from its ashes. Let flame consume
the perfumed gorse in all its glory and not spare the broom,
whose more exquisite yellow atones for its lack of fragrance.
In this common ruin be every lesser flower involved: blue beds
of speedwell by the wayfarer's path, the daintier milkwort and
rougher red rattle, down to the very dodder that clasps the
heather; let them perish and the face of Dame Nature be
utterly blackened. Then shave the heath as bare as the back of
your hand, and if you have felled every tree, and left not so
much as a tussock of grass or a scarlet toadstool to break the
force of the winds, then shall the winds come, from the east
and from the west, from the north and from the south,
and shall raise on your shaven heath clouds of sand that
would not disgrace a desert in the heart of Africa. By some
such recipe the ground was prepared for the Camp of
Construction.'

To sensitive country dwellers, however, the spoliation of nature was not as repugnant as were the hundreds of workers who were building The Camp, together with the masses of unsavoury men and women who squatted on its outskirts. The workers had trekked to the scene from all parts of the country, attracted by the record wages being paid there for building and construction work. Most came without womenfolk and children. They camped under canvas or in hurriedly constructed shanties. Then to live off these affluent workers and the thousands of troops there followed, mainly from London, a motley mob of traders, hucksters, beer and liquor sellers, gamblers and confidence men, entertainers, prostitutes and pimps. In addition there were many gypsies who had long wandered the lonely heath lands. The scene was often compared with the tough and rough shanty settlements of colonial Australia or with the lawless pioneer mining settlements of the American west. By day the workers toiled, while the troops marched and drilled. By night the ramshackle quarters of the traders, of the drink and vice vendors, became illuminated by oil lamps and flares. The thousands of soldiers and workers flocked towards these for their off-duty fun and recreation.

Many of the drink shops, put roughly together with canvas, wood or sheets of zinc, had a large area behind the bar separated by a curtain. Here there were rows of chairs and a platform where variety artists appeared. Comedians cracked dirty jokes and bawled dirty songs. Women and girls, often billed as 'songstresses' or 'chanteuses', shrieked popular songs and, in dancing, saucily flipped up their skirts to reveal their long lacy drawers. There were clog dancers, acrobats, conjurors, usually cavorting to the background music of a few brass instruments.

The area was to the crude labourer and soldier a sexual paradise, but too often leading to a sexual hell. Heavily perfumed and painted females rubbed their bodies against those of the males in the drink shops. Nearby were 'love tents' to which the prostitutes led the men who had been successfully solicited. But many poorly paid private soldiers preferred leading girls into the darkness and spaciousness of the heath lands beyond The Camp for the fleeting delight of what was called 'a quick 'un' or 'a short time' or 'a dip of the wick'. Sordid details of this commercialised sex were observed and recorded in secret reports by the few religious and social workers who at that time had the courage to penetrate into the area to listen and observe. Details of tragic and agonising sequels to many of the encounters between man and the female prostitute are preserved in the records of the now extinct Farnham workhouse which treated many syphilitic cases, a large proportion of which were fatal. Many previously respectable young girls of Farnham and the neighbouring villages were, of course, attracted to the 'glamour' of the Aldershot shanty settlement which teemed with so many men. Throughout the district there was a big increase in the bastardy rate; also of babies born into blindness, a legacy of gonorrhoea.

In general the queen, Albert and the army command, for all their probity, ignored the moral degradation which the creation of The Camp involved. They were sentimental about the ordinary British soldier and about the labouring man too. Yet it seemed that, at the backs of their minds, they also saw them as animals whose instincts could not be denied. Prostitutes flourished near every army centre without moral or hygienic checks. They were a recognised institution among the British army in India. We shall, in later pages, meet brave and idealistic men and women who were to come to Aldershot to provide soldiers with other diversions than sex and booze. At the start of The Camp, however, religious folk held up their hands in horror and protested without offering any attractive and sensible alternatives to bestiality. Many of these were poorly educated itinerant preachers, so common in the Victorian age. They carried banners and waved bibles, threatening the roistering troops with hell-fire. But the sinners preferred the beer

mug and the 'dip of the wick' to any consideration of their fate
in the hereafter.

The Prince Consort frequently visited The Camp which was
virtually his creation and he decided that it was good. Sometimes he slept under canvas there, but at night he did not,
unfortunately, inspect the area where thousands revelled in the
saturnalia of lust and liquor, although he must have been
disturbed by its noise. The prince focussed most of his attention
upon the rows of tents already occupied by the forces and upon
the stone or brick barracks under construction. He watched
officers and men closely. He pursed his lips in disapproval when
he thought they lacked smartness in their appearance and in
their drill. While in The Camp one day he was astounded
when a young officer approached him with a lounging gait,
dressed not in his regimental tunic but in a shooting jacket. On
the same day he jotted down a reminder to himself to complain
in high quarters regarding the shoddy appearance of the Rifle
Brigade. However, this friend of the army and creator of
Aldershot was often pained by the lack of appreciation shown
in some quarters for his intervention in military squabbles,
large and small. He somehow became involved in a ridiculous
controversy between commander-in-chief Lord Hardinge and
Adjutant General Sir George Brown over how much knap-
sacks ought to weigh. Newspapers told Albert to stop interfer-
ing.

The wandering evangelists had warned the labourers and
troops of The Camp that their sins would be punished by hell-
fire. But in December of 1854 retribution fell upon them in the
form of snow and ice, not flames. In this December there
began the coldest winter of the century. Many in and around
The Camp were stricken by pneumonia and frostbite. The
shanty shops and even many of the tented bars closed down.
The regiments of tatty prostitutes were depleted by the departure of many for London in pursuit of warmth. Snow and
ice prevented the moving of building materials. There was

such chaos that all construction work at The Camp was ordered to be suspended until the coming of a thaw. Many workers who had saved money sought lodgings in nearby Farnham, where their wild behaviour in its taverns scandalised the townsfolk. In The Camp itself soldiers were allowed to spend most of their time trying to keep warm in their tents or huts. However, in January 1855 two squadrons of cavalry were routed out to ride through the snow to Bristol to help suppress labour unrest. And the army was warned to prepare to tackle possible troubles in other parts of the country. The poor were complaining they could not keep warm because coal had risen to the unprecedented price of 2s. 3d. a hundredweight.

Early in February 1855 Lord Aberdeen resigned. The new prime minister was Lord Palmerston. Largely responsible for the fall of this favourite of Victoria and Albert was his shocking mismanagement of the Crimean War. Critics also pointed accusing fingers at Lord Hardinge for the lack of military preparation for the conflict and for inefficient army organisation and training while it progressed. However, Hardinge remained commander-in-chief and began to act more vigorously under the spurs of strong-willed and energetic Lord Palmerston. The effects were soon felt in Aldershot by the appointment of Albert's friend Lieutenant-General Sir William Knollys as the first commander of The Camp. Knollys threw himself into his new duties with almost fanatical fervour, at the same time moving under a perpetual bombardment of suggestions from Albert. At the start the military organisation of The Camp had been non-existent. Every army unit there seemed to do what it pleased. Now Knollys brought all the troops together in brigades and divisions. He caused a wave of resentment among officers by suggesting that they spend less time on leave, attending balls in London and hunting in the shires. He organised departments for transport, food and medicine. 'Who is controlling our chaplains?' he inquired, soon afterwards merging them into a department too. Such deeds of the commander-in-chief spoiled the happy-go-lucky atmosphere of the fast-expanding camp so that many of the officers there resented him. They were supported by a section of the London press which sourly hinted that the Prince Consort was really responsible for the reforms which Knollys was pursuing. Some of the critics of Knollys thought the limit had been reached when he was seen crawling about the ground of The Camp presuming to show officers and men how to pitch their tents. Knollys would himself at times sleep in a tent to share the simple life of the troops. And on some of his numerous visits Albert could not resist the temptation of also sleeping in a tent.

It became obvious that, with the growing importance of The Camp, there must be dignified quarters where Queen Victoria could sleep when she came to Aldershot to visit her beloved army. Thus the Prince Consort, Lord Hardinge and other high officers spent a long time examining the terrain for a suitable site for the erection of an abode which would be known as the Queen's Pavilion. Eventually the ideal spot was selected. This was an area of raised land overlooking Long Valley and the prehistoric earthworks known as Caesar's Camp. This would also provide a clear view of the rows of barracks which were being planned. The site for the Queen's Pavilion was claimed as her preserve by one of Lord Hardinge's young officers who dropped sticks all around it from a horse.

The building of the Queen's Pavilion was begun early in June 1855. Ground on the hill was flattened by an army of manual workers before construction could start. Near the middle of June the queen came to Aldershot for her first inspection of The Camp. The Pavilion was making good progress. The queen, her husband and Sir William Knollys had a meal there. In July she was in The Camp again. She bustled about the Pavilion, now approaching completion, discussing ideas for its interior decoration with Albert and ladies-in-waiting. The court regarded the building which was almost all of wood—as merely temporary. Yet actually it was to remain to be occupied in turn by Victoria's successors Edward VII and George V and to be renamed the Royal Pavilion.

A member of the staff of the *Illustrated London News* inspected the Queen's Pavilion when it was finished. He described it as 'bald, cold and ugly to the extreme', adding that its walls and ceilings were made of 'canvas stretched on frames and papered over'. He remarked that the view from the windows was dreary. But at that time there were no trees around the Pavilion and the landscape was disfigured by uncompleted buildings and piles of building materials. The Queen's Pavilion had a breakfast-room, a drawing-room and a large dining-room with salon. There were four large bedrooms with dressing-rooms. Two of these were for Victoria and Albert, and the other two for important guests. Two wings of other rooms were for the queen's attendants. Every door, except those of Victoria and Albert, had small frames for the insertion of the names of the room's occupant. Standing separate from the main house was a building for the accommodation of seventeen royal servants. This had two dormitories—one for females and the other for males. The servants slept on diminutive camp beds and had washbasins above which were fixed small mirrors. In the basement was a large kitchen. Meals were conveyed from this kitchen to the Queen's Pavilion through a tunnel with a glass roof. At the Pavilion the dishes were placed on a hand-operated lift and hauled up to the dining-room. The remoteness of the kitchen and the long line of communication to the dining-room were due to the hatred possessed by the queen for the smell of cooking. Once the food was on the table she enjoyed the sight of it and often ate heartily. But the savour of meat and puddings in preparation made her feel sick.

The rooms of the Pavilion became furnished with a mass of bric-à-brac, some of it massive and more suited to Windsor Castle than to this super-hut in a camp. It was said, however, that whenever the queen was in residence she could, in such decorative surroundings, feel comfortably and cosily at home, particularly in the drawing-room. This was a long room with walls covered with pictures, mainly portraits. It had several long sofas upholstered in regency stripes. There were wooden 'camp chairs' with seats upholstered for regal comfort. Down the centre of the room were several mahogany tables covered by dark red tablecloths with tassels at their corners. At one end of the room was a massive grand piano. There was a front terrace (patrolled by sentries) where the queen could sit and enjoy a panoramic view of The Camp and its activities. The terrace had a small artificial pond, populated by goldfish, close to the Pavilion's front door. Nearby were two large and well-furnished huts for the accommodation of the army chief of staff and Minister for War on their visits to Aldershot. The Pavilion had its own stables for forty horses. Around the residence was planted a garden of heather and flowering shrubs. This was designed by Albert.

Since time hallows so many things it is not improper to comment upon the lavatories installed at the Royal Pavilion for the exclusive use of Victoria and her consort. The apparatus was, for the Victorian age, ultra modern, and their mass of shining brass fixtures was quite magnificent and out of keeping with the drab simplicity of much of the rest of the house. The large toilet seats were of glossy unblemished South American mahogany. When, in the 1950s, the Pavilion was dismantled and demolished for the erection of an army nurses' training college there was earnest official discussion on what should be done with these splendid former seats of the mighty. Eventually they were removed from their brass hinges and discreetly hurried into a plain Department of Works van. Their destination was said to be Windsor Castle.

On April 26, 1856, the queen and her consort dined and spent their first night in the Pavilion. They expressed themselves delighted with its cosy simplicity. In the morning they watched a mimic battle involving thousands of troops with much firing of dummy ammunition. Then the sky clouded over and heavy rain seemed imminent; so, for the convenience and comfort of her majesty General Knollys ordered the battle to cease forthwith. She and Albert returned to London from Farnborough railway station. But the queen was impatient to be back to enjoy the novelty of living in a camp. An opportunity came in the following July when into Aldershot poured a large number of troops just back from the Crimean War that had recently ended. It was decided that there should be a royal inspection at which the queen would formally welcome her soldiers home. On this occasion Victoria and Albert stayed for a week, and for the first time they entertained others in the Pavilion. There was Edward Prince of Wales, a pouting adolescent of fourteen, subject to lightning outbursts of ill-temper. There was the queen's favourite uncle and perpetual adviser Leopold I, King of the Belgians. Among those also present were the queen's cousin George, Duke of Cambridge, Prince Oscar of Sweden, the Count of Flanders. With commander-in-chief Lord Hardinge, recently made a field marshal, came the Minister for War Lord Panmure, each occupying an individual residential hut for the first time.

The field day, with inspection, was filled with pageantry, emotion and drama, but the unfriendly English rain robbed the proceedings of perfection. The queen rode out to the parade ground in her carriage. Around her on horseback were six distinguished escorts all in uniform. These were Prince Albert, the Prince of Wales, King Leopold, the Count of Flanders, the Duke of Cambridge and Lord Panmure. The regiments from the Crimea were formed up in a large square about the queen's carriage. As in 1856 any public address system was unknown, except for the megaphone, representatives of the regiments—officers and privates—marched to the carriage to receive a verbal message to pass on to all their comrades who were out of earshot. The queen said: 'I wish personally to convey through you to the regiments assembled here this day my hearty welcome on their return to England in health and full efficiency. Say to them that I have watched anxiously over the difficulty and hardships which they have so nobly borne, that I have mourned with deep sorrow for the brave men who have fallen in their country's cause, and that I have felt proud of their valour which, with their gallant allies, they have displayed on every field. I thank God that your dangers are over, while the glory of your noble deeds remain.'

After concluding with a few additional words the queen seemed shaken by the emotion of the occasion. There was a sharp, tense silence. Then from the men came the cry of 'God save the queen!' followed by roars of cheers which spread to all the thousands of troops in the distance. Swords were waved. Bearskins and helmets were tossed in the air. Afterwards the Duke of Cambridge described the scene as 'beautiful'. Next came a grand 'march past', but this was marred by drenching rain.

It was during this historic day that army commander-in-chief Lord Hardinge was, through an eccentricity of fate, making his last appearance on an Aldershot parade ground. The seventy-one-year-old field marshal was talking politely with the queen and Albert when suddenly he fell to the ground. He was helped to his feet. The queen was not too solicitous, believing that he had merely slipped and would not appreciate much attention being given to a minor mishap. Always the courtly gentleman, Hardinge asked Victoria's pardon for 'making such a disturbance'. Actually, as it was found later, the commander-in-chief had suffered a stroke, but had somehow managed to keep his mind sufficiently clear to apologise to his queen. His right side had become paralysed. Within forty-eight hours he had resigned, to suffer in retirement until his death in the following September. Once he had been a glittering national hero, but his blundering and inefficient administration of the army, both before and during the Crimean War, had rubbed off the gloss. Obviously too old and doddery for such high command, he was constantly and often cruelly being criticised by the newspapers. Indeed, the indignant queen insisted that Hardinge had been 'killed by the press'. Soon it was announced that the new commander-in-chief was Victoria's cousin George, Duke of Cambridge.

General Sir William Knollys remained as Aldershot's G.O.C. until 1860 when he was succeeded by General Sir John Pennefather. During this period The Camp and Aldershot town grew at a constantly accelerating pace. In the area of The Camp there appeared amenities that included three churches, the Prince Consort Library (presented by Albert), Victoria Soldiers' Library, an officers' club, two riding schools, a musical society and a theatre. Also founded were a fire station and a commodious cemetery. The theatre, named the Royal, was a wooden building to accommodate an audience of 500. The orchestra stalls were cushioned in thick plush. The theatre opened in 1856 with a play called *The Printer's Devil*, produced and presented by officers of the Royal Worcestershire Militia. The theatre was for amateur theatricals. Nevertheless it was felt that, although it was seemly for officers to appear on its stage, the presence of their wives and daughters there would be indelicate. Thus after much discussion it was decided that female roles in Theatre Royal should be played by professional actresses from London—but only females of unsullied reputation were employed.

The town of Aldershot grew as fast as The Camp. For an increasing number of ambitious persons from all parts of the country were coming to open shops and other businesses. Private builders, seizing the opportunity for fast and fat profits, bought up plots of land for the building of shops and houses. Thus such thoroughfares as High Street, Wellington Street, Victoria Street and Union Street were in evolution. There was an invasion of military and civil tailors, of butchers, grocers, ironmongers and stationers, hotel and lodging-house keepers. The return of regiments from the Crimea still further expedited the growth and prosperity of this boom town. The men and their officers had during their service abroad accumulated large arrears in pay. Now they were splurging it. Between 1851 and 1854 the civilian population of Aldershot town increased from 900 to more than 3,000. In 1861 it was nearly 8,000.

The town's most profitable activity was that of satisfying the army's thirst for alcohol. In 1851 Aldershot had two small public houses. In 1855—only a year after the foundation of The Camp—there were counted twenty taverns and forty beer houses. And many of these provided facilities and women for satisfying sexual thirst also. These new taverns and beer shops were additional to those on the rim of The Camp, already described. Hygiene and sanitation lagged far behind in Aldershot's race towards development. The place was a horror of filth, muck and stench. This was because the speculative builders were not inhibited by any compulsory sanitary regulations. No such regulations existed. The builders economised by not providing cesspools or even space for them. Townsfolk dug holes in any available ground for the reception of excrement and other waste, and any quiet wall or gutter served as a urinal. When the rains came the filth usually rose from its holes to become a kind of Hampshire River Styx. It encompassed the town and adhered to the footwear of its inhabitants. Officers from The Camp were nauseated by the sights and the smells, warning the G.O.C. of the health perils faced by the troops in such conditions. Many indeed did contract fever from their trips into Aldershot town. Civilian workers with sufficient means flocked into more civilised and cleaner Farnham to seek houses or lodgings. And this old town, also benefiting from The Camp, became notorious for its extortionate prices. In Farnham too there was a great expansion of inns and taverns.

Thus were Aldershot and The Camp conceived and born.

2

Dare to be a Daniell

Wellington, in his famous tribute to his troops of the Peninsular War, said: 'They are the scum of the earth, and it is really wonderful we should have made them the fine fellows they are. With such an army we can go anywhere and do anything.' The Peninsular War ended in 1814, but far into the nineteenth century the British soldier, while hailed as unconquerable, continued to be regarded as scum. Queen Victoria gushed tears over his loyalty and steadfastness. Tennyson and others wrote noble poems about him. In the grandeur of his dress uniform, his marching and his bands, he was cheered by the watching mobs. Yet off parade and in the confines of his camp or barracks he lived in conditions of squalid indignity. And although the rank-and-file were admired for their martial achievements, their colourful uniforms and accoutrements, their place in the British social scale was at the bottom. The army remained for a long time the refuge of youths and men who in civilian life were regarded as unemployable, for the illiterate, the ne'er-do-wells and the paupers. When former domestic servant William Robertson (later Field Marshal Sir William Robertson, Bart.) decided to enlist as a trooper in 1877 his disgusted mother wrote to him: 'The army is a refuge for all idle people. I would rather bury you than see you in a red coat.'

The recruit, once in the army, was savagely drilled and bullied into submission by the N.C.O.s, loyal to the traditions of cruelty, misnamed discipline, in which they themselves had been trained. The commissioned officers also believed that to be made into a good soldier the recruit had to be 'broken' like a wild and recalcitrant horse. Indeed, the soldiers were driven so hard and lived in such unhealthy conditions that for years the death rate in the army was five times that of civilians. Aldershot boasted about her modern huts and barracks, but most were damp, draughty and shockingly overcrowded. The rows of camp beds were so close that the breath of the sleepers intermingled. Smelly oil lamps provided a glimmering light until progress replaced them with gas jets. In really cold weather coal in the centre of the huts were allowed to be lit. There were set apart 'ablution huts' with rows of basins and pitchers of cold water where the men could wash. There were primitive latrines and, outside huts and barrack rooms, or in their corners, tubs were placed at night for the convenience of soldiers wanting to urinate or to relieve their bowels. For the soldiers' hours of recreation there were some sparsely furnished reading rooms, and school rooms for study.

The Camp had large cookhouses constructed of iron. The ration for the soldier was for many years twelve ounces of meat and a pound of bread a day. The meat ration included bone, fat and offal. There were also potatoes in plenty. The men breakfasted on bread and coffee, they dined at midday on meat and potatoes, with a bonus of suet pudding or soup two or three times a week. The final official meal was tea and bread. Incredible, but until 1870 the soldier was charged sixpence a day for his food from his shilling a day as an infantryman or 1s. 2d. a day as a cavalryman.

In 1871 and 1872 came what are known as 'the Cardwell reforms'. These, introduced by Lord Cardwell, Minister for War in the Gladstone cabinet, are often said to have revolutionised army life and to have at last made it attractive to the recruit. The compulsory period of enlistment had been twenty-one years. Cardwell shortened this to twelve years of active service and nine in the reserve. In a further reform the recruit could serve seven years with the colours and be on the

C

reserve for five years. Cardwell also abolished the purchase of
commissions. It was claimed that the army then attracted a
better class of recruit and more dedicated officers.

However, when the future Field Marshal Sir William
Robertson defied his scandalised working-class mother and
became a trooper in the 16th Lancers in the late 1870s he found
conditions appalling. He had been a footman, but he discovered
that as a private soldier he suffered a big fall in standards of
living and comfort. The regiments were, he wrote, 'mainly
composed of old soldiers who were in many cases addicted to
rough behaviour, heavy drinking and hard swearing'. For a
long time young Robertson was so miserable that he had to
struggle against a recurring temptation to desert. When on one
occasion he was about to flee Aldershot a comrade stole his only
civilian suit and made a getaway in it instead. Sixteen slept in
Robertson's small barrack room. Each bed contained a mat-
tress, pillow, four blankets and two sheets. The sheets seemed
a luxury, except that they could be changed only once a month.
New straw for the mattresses was provided every three months.
Usually the blankets, brown in colour, had to be filthy before
they were washed. No wonder that The Camp was also a
centre for brigades of fleas and pubescent lice. Barrack rooms
were furnished with plain wooden benches and tables. Soldiers
had to provide their own crockery which consisted of one
plate and basin a man. The basin served as a receptacle for
soup, tea and coffee, beer and shaving water.

Cruelty in the early Victorian army had majority official
support because of the widespread belief that the average
recruit was too low and too stupid to respond to kindness and
reason. Even small breaches of discipline could be punished by
savage flogging. In 1835 and 1836 there was a Royal Commission
on flogging when the Duke of Wellington in his renowned
wisdom stiffly opposed the views of commander-in-chief
Lord Hill, who, with surprising humanity, wanted to abolish
the lash. The Duke, who had once sentenced a thieving

soldier to 999 strokes of the cat, told the commission that if flogging were banned all discipline would disappear from the army. Other high officers strongly endorsed the views of Wellington.

In Aldershot's early years the chief punishment, as at other military centres, was 'pack drill'. This was imposed with such ferocity and for such prolonged periods that its victims sometimes died from exhaustion. There was also vicious and sadistic flogging for offences which today would be regarded as trivial. In July 1859 a correspondent of Charles Dickens's magazine *All The Year Round* visited The Camp. He wrote that if a soldier at the end of his evening leave 'neglects to return to his disconsolate regiment at the appointed period, he suffers for it the next day, and several following days, by the extra exertion of "pack drill", if not by a more severe punishment, for the shadow of the hateful "cat" still hovers over the pet military settlement, still comes up through the dust and theatrical glory of a sham field-day, still dims the brightness of the medal and the cross'. There could be heavy punishment for soldiers caught smoking or 'failing to keep in step' while walking in Aldershot streets. For desertion the punishment of branding—introduced in 1840—still lingered. The recaptured deserter was usually first flogged in the presence of his comrades. Next the letter D was branded on his quivering body with a device of needles. Gunpowder was massaged into the bleeding flesh to ensure that the D would remain indelible.

Visitors to The Camp used to be surprised by the simplicity of the hut or barrack accommodation for officers. The subaltern slept in a small, plain room. Its furniture consisted of an iron bedstead, washstand and two chairs. The room had a fireplace and an oil lamp. But its smallness and furniture left the occupier little space for movement. According to the correspondent of *All The Year Round* 'there was about a square yard of flooring for exercise and toilet'. The writer added that the officers' rooms 'look like the lodgings let to single men about Stepney, at two shillings a week, or the summer-houses that used to be erected in the grounds of market gardeners at Hoxton'. But the officers had comfortable and relatively commodious mess halls and from 1858 they had their own club in The Camp, with armchairs, billiards and other amenities of the fashionable clubs of London.

Anyhow, the smallness of the residential huts was not an overwhelming inconvenience to the well-heeled officers, many of whom had purchased their commissions. They enjoyed long periods of leave when they could recuperate in their mansions and country houses. Then, except for special parades and manœuvres, most of the strenuous work in The Camp was left to the N.C.O.s. Indeed, officers would not normally appear for duty until after eleven o'clock in the morning. Although the young subaltern or captain was not expected to be a diligent officer, it was obligatory that he should be a diligent gentleman. The army command was particularly insistent that he should look the gentleman while off duty. Before the young officer left for trips to London his superiors looked him over to ensure that no vulgarity such as too loud a tie or too bold a tiepin made him appear too common for one holding a commission in the queen's army. If, for example, a superior officer chanced to see him in loud or otherwise inappropriate attire in a London club he received a stern reprimand. For no officer was allowed to emulate the appearance of 'bounders' and 'cads'. Most would even travel from Aldershot to town in frock-coats and top hats.

To revert to Aldershot's men of the ranks, their burdens were painfully increased if they were foolish enough to fall in love and to sanctify that love in legal marriage. No one worried much if the soldiers confined their passions to the whores who flourished so plenteously in Aldershot and the surrounding heath lands. Indeed, as has been emphasised, the 'dip of the wick' or the 'blow through' was considered necessary to the physical comfort of the men despite the constant risk of venereal disease. Legal marriage, however, encumbered the army with wives and children who complicated the life of The Camp and for whom there was inadequate space or financial provision. This provision, tiny as it was, was made for only a limited number of couples in each regiment. The wives admitted into this favoured category were known as 'on the strength'. They lived in The Camp with their husbands and contributed towards their keep by doing laundering, mending and other domestic jobs for the regiment. They and the children accompanied their husbands if the regiment was moved to billets in other parts of the British Isles or even to some garrisons abroad. When it was not possible for 'on the strength' wives and families to accompany their menfolk overseas they received a small maintenance allowance.

Rules and regulations, however, prevented the regiments from becoming encumbered with too many wives or children. There was a strict quota limiting the number of wives 'on the strength'. Thus a large proportion of the soldiers who married had to keep their wives and children 'off the strength' in homes away from The Camp and without army grants towards their food and clothing. Since the soldier's pay was so pitifully small, these unfortunate women and their offspring existed in ugly poverty. However, in its compassion, the army command did not object if a married soldier denied himself his full ration of food to carry the remainder to his hungry family.

Many written descriptions exist of those privileged wives and families who lived 'on the strength' in The Camp at Aldershot and at other military centres. Before 1854 heralded the progressive wonder of Aldershot the married men, their wives and families, occupied the same rooms as the bachelor soldiers in some barracks. The 'married quarters' in these rooms would be separated from vulgar bachelor gaze by screens of grubby tattered blankets or sheets. Of course there could be no privacy of sound, so that at night creaking beds and other noises from behind the curtains would result in ribald laughter and remarks from listeners. A commission investigating such accommoda-

tions in 1855 demanded better and more decorous conditions
in barracks and camps. Thus in each regimental block in
Aldershot one hut or room was reserved for the occupation of
married soldiers and their families exclusively. In each such
connubial hut or room there would be from eight to ten men
with their wives and children. Each family endeavoured to
preserve its privacy at night by the screens of blankets, sheets or
wet washing around their beds. From early morning onwards
married quarters could quickly be identified by the shrill cries
of females, the screams and laughter of a horde of children and
babies. In the middle of each hut or room was a communal
stove to be shared for cooking by all the resident families.
Pervading the quarters was the smell of food intermingled with
that of human sweat and drying washing. There were also
special reading and writing schools for the 'on the strength'
children.

In her book *Aldershot and All About It* Mrs. Young, a
Hampshire writer, described her entry into a hut where
resided ten soldiers, their wives and eighteen children. She
wrote that after 'a night of jarring discord' the residents
'hastened to raise the ragged sheets or rugs which form the only
screen between the beds, and hurry forth the little ones, bread
and butter in hand, to the infant school'. Next the authoress
saw the women put their saucepans on the centre stove. The
floor of the hut, she added, was 'littered with bones and potato
peelings', and 'men's washing dripped from lines overhead'.
Visitors to The Camp who saw its children were impressed by
their robustness and happiness. The boys and girls enjoyed the
freedom of all parts of The Camp. They were adored and
fussed over by the hundreds of soldiers with whom they
mingled. Their mothers were described as wild and slatternly,
and addicted to brawling and fighting among themselves in
quarrels over the inadequate cooking facilities and over the
ownership of garments and blankets in the overcrowded
communal quarters. Nevertheless they must have been brave

women to endure the hard and unnatural life to which their marriages condemned them.

Outside The Camp and near the wall forming the boundary of the South Cavalry Barracks was a district of Aldershot known as West End. Here in irregular rows of shoddily built little houses dwelt most of the wives of soldiers who had wed without official permission. They were therefore 'off the strength' and as such were denied any maintenance allowance or accommodation from the army. Since the pay of the privates was so small, these wives and their children were desperately poor. The frocks of the women were often ragged and tattered. Many of the boys and girls, and even babies, wore garments clumsily converted from the discarded uniforms and underwear of their fathers. Few of the dwellers in West End had sufficient money to buy the food for which their starved appetites craved. However, most were sustained by food that the soldier fathers had managed to save from their own meals in The Camp and to scrounge from their comrades in the regimental cookhouses. These women and girls of West End sacrificed so much to romance in defiantly wedding such poor men that their physical looks and general appearance would quickly and sadly deteriorate. They were shunned by many Aldershot townsfolk who regarded themselves as greatly superior to such degraded females. The wives of soldiers 'on the strength' enjoyed far higher status in and outside of The Camp than the 'off the strength' wives of West End.

Aldershot's West End was denigrated as being low and sordid. Yet that district of poverty and bad fortune was nothing so terrible as developments in the early 1860s spread in the centre of the town as insidiously as the infectious disease they encouraged. In roads off the High Street and elsewhere were established a mass of drinking places, dance halls and brothels. The shanty town and squatters' market on the heath land bordering The Camp, with their entertainment, booze and vice, declined. Now such squalid recreations were offered to the troops in more permanent and spacious buildings. These new businesses were established by additional newcomers from London and by men and women who had prospered from the custom of soldiers and labourers in the shanty area. By day Aldershot seemed provincial and respectable, despite the proliferation of public houses on and bordering its main streets. But by night it became a vulgar and boisterous fun town, shunned by all respectable women and girls. Drunks rolled about its smelly streets. Oily pimps sidled up to young privates to tell them in grinning whispers where they could find 'cheap and clean' girls. And from the doorways the prostitutes themselves shouted at passing males and beckoned. Moreover by night the town was infested by ruffians who would often use women decoys to entice men into dark corners or passages to be attacked and robbed.

Booze and girls were served to the poorly paid privates at bargain rates. Strong beer was available at a penny a pint and glasses of adulterated spirits at a little more. A girl could be enjoyed for a few minutes for sixpence, or even less when there was a scarcity of customers. Hardly an evening passed without picquets from The Camp charging in to break up drunken brawls involving troops. The command of The Camp, while still feeling that troops must be allowed plenty of licence during their nights of recreation, at last became shocked by conditions. A report about them was prepared by a Captain Alkington-Jackson. The captain accused Aldershot townsfolk of encouraging or acquiescing to indulgence in alcohol and vice which gravely threatened the health and welfare of the troops. But the reaction of Aldershot people to this report was to hold a protest meeting hypocritically to allege that their fair town had been foully libelled. Apparently there were only a few civilians who were not profiting from the shillings and pennies squandered in self-indulgence by thousands of soldiers on nightly excursions into town.

During most of the 1860s Aldershot had inadequate facilities for maintaining law and order. The police were few and they preferred to shun the districts of vice and violence after nightfall. The local court was held in the ballroom of an hotel by the chief magistrate Captain Newcome, one of the scarce local gentry. With a better understanding of foxhounds than of men, Newcome concealed his ignorance of the law by bawling and banging a gavel at those brought before him, whether guilty or innocent. Next, doctors began to warn that venereal disease was spreading fast among the civilian youth of the town as well as among the troops. Yet nothing was done to shut down the brothels or to reduce the drinking shops and the dance halls. But a Lock hospital was established on Greenham Hill, Aldershot, while the army started to send many of its worst syphilitics to the new military hospital at Netley, near Southampton. Almost simultaneously it was revealed that panders from the town were visiting The Camp itself to mingle with the troops and drum up even more business for the brothels.

The youths and young men of the army were obviously in dire need of genuine friends, of moral guidance. They deserved recreational alternatives to drink and women and dancing when they went into Aldershot. In the early 1860s there were four Anglican army chaplains, two Roman Catholic priests and one Presbyterian minister trying to minister to the men of The Camp. But as the number of soldiers often exceeded 25,000 they had an impossible task despite additional help from four scripture readers representing the Soldiers' Friend Society. The only time most of the troops saw the chaplains was when they marched into garrison churches at formal military parades on Sundays. In the town, too, religious and social facilities were inadequate to meet the problems created by vice and alcohol. There were pathetically few counter-attractions for those soldiers and others who were being dragged into evil. In 1860 there opened in Aldershot the Institution for Mental Improvement and Social Recreation. Because the institution coldly concentrated upon 'mental improvement' through study and lectures and provided few recreational attractions, it failed to draw the hundreds of young men who nightly came into town for pleasure. The local clergy denounced the drinking and the vice, but they were too few in number to exert real influence. However, there were signs of growing religious awareness. The ancient parish church of St. Michael was enlarged in the 1860s. A Presbyterian church was built and a Roman Catholic mission was opened. A synagogue was also established for Jewish traders and soldiers. But in these developments there seemed to be more emphasis on bricks and mortar than on saving men from venereal disease and alcoholism.

However, in 1862 a saviour and crusader arrived in Aldershot. She was a plain little woman wearing a widow's bonnet, with her dark hair parted in the middle and strained severely into a bun. The woman was Louisa Daniell, aged fifty-three, whose late husband Frederick Daniell had been a captain of the 1st Madras Native Infantry. Louisa Daniell was a woman of earnest piety who, wherever she went, tried to convert people to her own individual brand of extreme protestantism. It was while she was staying in a Midlands village, recuperating from illness, that she became obsessed with the idea that many lowly country folk, neglected by the established churches, were in desperate need of conversion and salvation. This obsession led her to start evangelistic missions in villages. In these she distributed tracts of her own authorship, held prayer and exhortation meetings in cottages and tried to alleviate poverty and suffering among the villagers. She was such an uncompromising protestant that she campaigned against the Roman Catholic Church. She struggled to prevent country folk from becoming what she called 'seduced' by priestly missioners and nuns who to her were agents of the devil.

Despite the excessive sanctity of Mrs. Daniell and her anti-Catholicism and other prejudices she did marvellous work in improving the moral, spiritual and even the material welfare of

hundreds of common country folk. She gave them a powerful
new faith from which sprung self-confidence and pride. In all
the villages where Mrs. Daniell laboured immorality notice-
ably declined, together with poaching, theft and other common misdemeanours. And, on the other side, Mrs. Daniell
succeeded in persuading landowners to treat their humble
tenants and workers with more sympathy and generosity. This
officer's widow, in her burning religious zeal, was said to
possess an uncanny power of persuasion which neither the rich
nor the poor could resist.

A male member of an organisation called the Country
Towns Mission visited Aldershot in 1861. He was appalled by
what he saw and became almost convinced by the defeatist
opinion of church people there that nobody and nothing could
snatch thousands of soldiers from a condition of degradation
which in them seemed innate and beyond alleviation. Nevertheless he thought of Louisa Daniell, who was associated with
his country mission, and of her almost mystical power of
influencing others. Thereupon he sat down and wrote her a
letter beginning with the words, 'I wish you would adopt
Aldershot.' In a later comment about this invitation Mrs.
Daniell said, 'The idea of adopting Aldershot seemed so
strange that I laughed at the suggestion.' But soon 'an inward
voice was saying, "That is the work for which you have been
praying." ' Mrs. Daniell made further inquiries about Aldershot
and was given 'the loathsome details of the unblushing vice
that tracks the everyday life of the poor soldier'. And she was
told of an army officer's evidence that 'nothing ever said of the
abounding wickedness of Aldershot could go beyond reality'.
Mrs. Daniell also learned that in the town of Aldershot 'there
are 100 public houses, some with dancing saloons and other
arrangements, by which these wicked panderers to vice entrap
the unwary; and the moment the poor soldier, tired with the
forced inaction of camp life, sets his foot beyond the lines, his
case is desperate'.

Of course in our much vaunted permissive society such anxiety over the temptations endured by 'the poor soldier' seems too sanctimonious and hypocritical. This is only because the Victorian sense of propriety forbade public mention of the real cause of the anxiety—the horrors of venereal disease. In those days there were no penicillin or other powerful antibiotic drugs to strangle the disease at its inception. The pimps who infested Aldershot and The Camp offering youths 'cheap girls' were simultaneously offering them disfigurement and probably slow death. Stories have survived of visits paid by mothers to the hospital bedsides of their venereal diseased soldier sons. On seeing the faces of these youths, eaten into hideousness by the disease, mothers would cover their eyes and some would run away screaming. Mrs. Daniell in her religious talks often reminded her hearers that 'man was made in the image of God'. She was well aware of the actual disfigurement of many of those young images whom no one had seriously tried to warn and to save.

Mrs. Louisa Daniell moved to Aldershot. She was introduced into The Camp to small groups of officers and men who, in defiance of convention, disassociated themselves from the immorality of the town and endeavoured to follow Christian ideals. She exhorted them; she prayed with them; she implored them to rally about her in the organisation of a mission. 'I have been asked and the Lord has laid it on my very heart to do something for the soldiers here,' she told one gathering, 'but we want your help, and we know it was never said that a lady appealed to a British soldier for help in vain. Will you come forward and help us? For we cannot do without you, and the Lord hath need of you. Let me press on the exhortation of St. Paul, "I beseech you, therefore, by the mercies of God, that ye *present* your bodies a living sacrifice, holy, acceptable to God, which is reasonable service." ' Surely this was neither a striking nor original appeal. Yet, responding to Mrs. Daniell's psychic charm, men surged around her offering to do all in their power to make the mission a success. Then Aldershot's protestant clergy and the chaplains of The Camp met Mrs. Daniell. Boldly she said that while she wanted 'to work with and not against' them, her mission must be independent and she could not place it under their control. They thought this declaration unwomanly and unwise. Nevertheless they agreed to give her their support.

Mrs. Daniell leased a house as a temporary mission hall and reading room. This was an immediate success. Soldiers who abhorred the church parades and who shunned services at the dignified parish church flocked to this hall to roar hymns, listen to Mrs. Daniell's strangely attractive sermons and patiently to endure her long extempore prayers. She was often assisted in her administrations by a number of 'gentlemen of proven piety' led by Lord Radstock, a Mr. Baxter and a Captain Fishbourne. However, the mission hall was merely a tiny start. Commenting in her picturesque prose on her arrival in Aldershot, Mrs. Daniell had said that 'the first glance of the low public houses and dancing saloons told us that the great want of the place was a *public* house opened for different purposes and conducted on totally different principles where both soldiers and civilians might be invited to pass those seasons of leisure now spent in the pursuit of pleasures which only debased the mind and hardened the heart'. By February 11, 1863, Mrs. Daniell's ambition for her unique *public* house was in sight of realisation. For then, on a plot of land presented to her, the Earl of Shaftesbury, prominent Christian and social reformer, laid the foundation stone of what he officially named the Soldiers' Home and Institute.

In his address at the foundation stone Lord Shaftesbury challenged the widely accepted belief that soldiers could never be transformed into decent living Christians because they belonged to a class so hopeless that they were beyond hope of reform. He told of letters written by private soldiers during the Indian Mutiny and Crimean War, revealing a deep religious faith. He also referred to British army officers—'Christian soldiers whose names shone so brightly in the pages of history.' But the officer class needed no defence from Lord Shaftesbury. No one doubted that officers, unlike mere privates, could not as born gentlemen be entirely incorrigible. The ceremony of laying the foundation stone was not without indecorum. For men and women from Aldershot's nearby streets of vice were looking on, jeering and laughing at the pious Lord Shaftesbury and plain, drab Mrs. Daniell in her widow's bonnet. There had during the winter been a spectacular increase in the town's low dance palaces. Some were visible from the scene of the ceremony, causing one soldier to comment to another: 'Just look at those dancing-halls. You could throw a stone over twenty of them.'

The Aldershot Soldiers' Home and Institute was opened on Sunday, October 11, 1865, and there followed a week of almost continuous rejoicing and prayer. It was a handsome if austere building of grey Kentish ragstone, with the inscription over the porch: 'Our God, we thank Thee, for all good things come of Thee, and of Thine own we have given Thee.' Inside on the ground floor were a large hall for meetings, a bar for non-alcoholic refreshments, a smoking and games room, a dining-room, library and reading room. Upstairs was a drawing-room where officers only were admitted, a classroom and other rooms. In the basement was a kitchen. Within a few years the premises were enlarged. The institute was a sensation, for never before had such a place for recreation been provided for the use of private soldiers and even drummer boys. Study and recreation huts in The Camp were furnished with bare tables and bare forms, often made more unfriendly by splinters. The institute contained stuffed armchairs, sofas and carpets. There were tasteful pictures on its walls in addition to a plethora of coloured texts. The bar in this Mrs. Daniell's version of a public house was opened daily except on Sundays. It served a large cup of tea for twopence and a small cup for a penny. A large coffee cost a penny and a small one a halfpenny. Cocoa was a penny, with a variety of cake at a penny a slab. In the dining-room for a few pence could be devoured soups, generous cuts from hot joints, vegetables, puddings and tarts. There was a strong feeling among old-fashioned officers of The Camp that the troops were being spoiled and that the institute was much too good for them.

The institute did not start operations without a crisis. This was caused by the problem of who should wait upon the soldiers in the bar and dining-room. Mrs. Daniell's idea was that such work should be undertaken on a voluntary basis by the wives and any adult daughters of those enlightened Christian officers who had helped her to establish her institute. She had also anticipated similar voluntary help from the womenfolk of the neighbourhood's sympathetic Christian gentry. It should be stressed that in seeking to minister to common soldiers in this fashion, Mrs. Daniell was trying to set a precedent. For in 1865 the institute was unique, only later to be followed by a torrent of wholesome social amenities for the troops by the churches, Salvation Army and other organisa-tions. Certainly Florence Nightingale and her female asso-ciates had given wounded soldiers intimate ministrations in the suffering of the Crimean War, and female nurses were essential in all hospitals. Nevertheless it was widely felt that for gentle-women to wait upon privates and N.C.O.s in time of peace was both unwise and improper.

Under such circumstances ladies would overhear the rough cursing, swearing and blasphemy; and their proximity to the men might invite undesirable familiarity. It was also felt that for officers' wives to serve privates might even affect army discipline, ending the strict social separation between officers and other ranks. Mrs. Daniell was persuaded—but only tempo-rarily. The assistance in her institute would, she decided, be provided not by gentlewomen but by women, such as respect-able wives of N.C.O.s and middle-aged female persons of strict moral probity from the town. This arrangement lasted until, gradually, recognition of the ordinary soldier as a human being rather than as a descendant of Wellington's 'scum of the earth', permitted officers' wives and other gentlewomen to work in the tea and coffee bar, in the dining-room and kitchen.

Mrs. Daniell had a co-worker in her enterprise who, in many ways, was as remarkable as herself. This was her devoted spinster daughter Georgina Fanny Shipley Daniell. She accompanied her mother to Aldershot and continued working in the institute twenty-three years after her mother's death in 1871. The two, indifferent to silly Victorian conventions regarding what ladies might and might not do, resolved from the foundation of the institute to make it truly a home from home for lonely and homesick soldiers. Mrs. Daniell would assume the role of their mother and Georgina that of their sister. They paid special attention to new raw and shy recruits, pouring out sympathy and advice to them in their troubles and fears. It seemed amazing at the time, but hundreds of young men forsook the drink and vice hells of the town to spend all their hours of recreation under the loving eyes of the two Daniells. Many became converted Christians and signed the pledge of total abstinence from alcohol. Before long, stories of the compelling power and influence of these two women became so widespread that mothers of youths who had just been posted to Aldershot would write to Mrs. Daniell begging her to keep an eye on them. Then Mrs. Daniell would enter The Camp, call at the barracks and invite the newcomers to make her institute their home from home. She would after that, in her own words, 'throw a living chord around them'—a chord of love and understanding.

Georgina Daniell kept and cherished many letters written to herself and to her mother by soldiers. These were full of gratitude for the happy, homely atmosphere of the institute. She also recorded remarks made to them by soldiers, such as, 'If you believe me, ma'm, it is like going out of hell into heaven to come up here from one of our barrack rooms.' And officers also wrote to the institute expressing thanks for its successful work among their men. An increasing number of officers also frequented the institute, but except for attendance at Mrs. Daniell's large religious meetings they spent most of their time in the drawing-room, out of bounds to other ranks. Often she would sit there talking with them, usually concluding with the suggestion: 'Shall we have a little prayer and bible reading before we separate?'

Writers of goody-goody Victorian novels and tracts often described heroic Christian men and women who suffered ridicule for their uncompromising faith. In The Camp this actually happened. Officers, converted by the Daniells, would rise in the mess and testify until gently asked to sit down by amused and embarrassed comrades. Privates, back in their hut or barracks from an institute meeting, would kneel at their beds in prayer while others, shouting obscene imprecations, made them the target of boots and containers of urine. These converts would proudly tell the Daniells of mockery and assault bravely endured by themselves and others. One described how in a tent on manœuvres a soldier knelt down in prayer while another grabbed 'a great piece of wood' and struck him on the head with it.

From these modern days the Daniells seem to have been painfully and unnaturally pious with their constant prayer meetings and bible readings. They also expected their followers unhesitatingly to exhibit their faith in face of most hostile and difficult circumstances. Nevertheless intermixed with their exhibitionist sanctity was a capacity for undertaking unpleasant practical tasks which most normal folk avoided. It was therefore natural to Mrs. Daniell and Georgina suddenly to turn their sympathetic eyes towards those women and children from whom others shrank, as if from lepers. These were the off-the-strength wives and families of soldiers living in squalor in Aldershot's West End. Details of their plight reached Mrs.

D

Daniell after a visit of missioners from the institute to West End. They went from house to house making notes of all they saw and heard. They found conditions as bad or even worse than in the slum hovels of the East End of London. In a report on the scandal Mrs. Daniell told how 'one poor woman was found only the other day with a new-born infant with nothing to cover it, and herself (poor creature!) with only one torn garment just over her shoulders, without nourishment, and owing a large sum for rent'.

Soon an invitation was circulated among the West End women to gather weekly at the institute for mothers' meetings and sewing parties. Tattered and bedraggled, they crowded into this hall, many with damp howling infants. Mrs. Daniell led them in prayers full of hope and optimism, and urged them to join heartily in the singing of the bright hymns. Still more practical, the women—most of them semi-starved—were served with free tea and buns. Mrs. Daniell told them that soon they would be able to help themselves by sewing garments and other articles for which she could find a market. Then the women attended needlework lessons and started sewing in their homes. The articles thus produced were immediately bought by Mrs. Daniell for resale to shops and other purchasers. This added from one to five shillings to the weekly income of these army wives.

Next, Mrs. Daniell and daughter set about to try to educate the wild, dirty brats of the West End who, she said, needed firm discipline even more than elementary knowledge. These efforts led to the leasing of a public house there called the Wellington Arms which also had a large dance hall with the entrance from the bar. The dance hall was turned into a classroom, but not until the uninhibited jokes, rhymes and drawings pencilled there by patrons and their girls had been obliterated. The shocked Georgina Daniell recorded how 'paint, whitewash, soap and water', were liberally applied to efface 'the indescribable filthiness which seemed like an outward expres-

sion of the moral corruption of the place. Sundry announce-
ments relative to stout, ale, porter and shaving were obliterated
from the doorposts.' The premises also served as a night school
for working boys aged between twelve and sixteen. They were
taught by three soldiers on evening leave from The Camp.

The biggest event of the year for the Aldershot Soldiers'
Home and Institute was its annual picnic. This was usually held
on the estate of one of the gentry in pastoral surroundings
several miles from town. Early in the morning some sixteen
wagons gathered near the institute while crowds of happy
soldiers, wives and families climbed on to them. While the
soldiers were uniformed, most of the wives had their worn
dresses covered by freshly washed white aprons or pinafores,
these being regarded as appropriate attire for working-class
females on gala occasions. The girls and smaller boys wore
pinafores too. Before the wagons started Mrs. Daniell said a
prayer. Then the picnickers moved towards the countryside,
singing hymns and patriotic songs. On reaching the estate they
settled on the grass, usually under a clump of trees on high
ground with a sweeping view of the countryside below.

Tea was served from giant canisters, and ginger beer from
stone magnums. Helpers bustled about with trays of cakes,
bread and butter and jam. Mrs. Daniell would watch the simple
happiness of her picnic with high pleasure. This was the only
'holiday' away from home in the year for most of the women
and children, and soldiers were granted special leave for the
excursion. The picnic ended with a short address and a hymn.
Then, in the dusk, the wagons returned to Aldershot. Many of
the picnickers would be clasping in their hands bunches of
wilting buttercups, cowslips and other wild flowers so abundant in the Hampshire and Surrey meadows before the intervention of the chemists with their weed-killers and insecticides.
Even Mrs. Daniell, in a rare concession to the vanities, would
pin a few fresh flowers in her widow's bonnet among the faded
artificial blooms already sewn there. Nothing could have been

so simple and so innocent as this picnic. Yet the poor wives and children—and the soldiers too—thought it wonderful. According to Miss Daniell, some cried when back in Aldershot because the picnic was over and because there would be a wait of twelve months until the next.

Mrs. Louisa Daniell was a good and marvellous woman; but she had her shortcomings. While full of understanding for young soldiers in their moral temptations, her attitude towards the whores with whom many had consorted was hard, uncompromising. Her abhorrence of these women and girls seemingly prevented her from displaying some compassion and trying to reform them as she reformed the soldiers. To her a fallen woman, like an axed tree, could never rise again. Thus while Mrs. Daniell toiled for the betterment and welfare of the troops and their womenfolk, she did nothing practical for those females whom male sensuality had degraded. She lacked the undiscriminating compassion of William Ewart Gladstone who would roam the streets of London by night, risking defamation and blackmail, to befriend the whores, those pathetic victims of Victorian heartlessness and hypocrisy.

During the years spent by Mrs. Daniell in Aldershot, the exploitation of females was scandalous and merciless. Teenage girls were enticed from lives of decency and even kidnapped for the brothels of London and the garrison towns. By hypocritically rigid Victorian standards the 'fallen woman' was seen by many as beyond redemption whereas in the words of social reformer Mrs. Josephine Butler 'the question of redemption is never raised' over the young man who regularly visited the brothels. He 'was merely supposed to be sowing his wild oats'. Mrs. Butler was also puzzled by the insistence of her contemporaries upon 'purity' in women, while failing to denounce those impure males who sullied that female 'purity' in commercialised vice. True that Mrs. Daniell sought to redeem the men while regarding the possibility of the redemption of the prostitute as hopeless. And when more optimistic Victorians did try to rehabilitate prostitutes, the latter were installed in institutions bearing tell-tale names as 'The Home of Mercy', 'Penitentiary House' and suchlike. One organisation to save prostitutes operated under the name of 'the Association for the Reclamation of Fallen Women'. Then, while expressing its repugnance and horror over the existence of the fallen ones, Victorian society paradoxically perpetuated the plight of many through Bills passed in 1864, 1866 and 1869 for the compulsory medical inspection of prostitutes. This measure had the effect of licensing hundreds of brothels. The Bills resulted not in a decrease in venereal disease, but a substantial increase. Yet the medical inspection order was not suspended until 1883. Mrs. Daniell's voice was never heard protesting against such evil follies.

Another weakness of Mrs. Daniell and daughter Georgina was their bitter and active hatred for Roman Catholicism. Roman Catholics were in the forefront in the United Kingdom in efforts for the reclamation of the 'fallen woman', working with admirable understanding and sympathy in a hard and merciless era. In the army Roman Catholics were numerous, particularly in the Irish regiments. Of course their faith helped them in their struggle to lead decent lives. The Daniells admitted Roman Catholics to their mission, but simultaneously talked against what they termed 'the dangerous romanising influences'. In the biased vocabulary of the Daniells people who became converted to Roman Catholicism were 'perverted to romanism'. They recognised no Roman Catholic converts, only 'romanistic perverts'. Dark corners of the otherwise illuminated Daniell minds were populated by dangerous Jesuitical priests and deceitful nuns bent upon robbing England of her protestant tradition, and luring soldiers to loyalty to the Pope rather than to the queen.

Louisa Daniell, so Victorian in her religion and piety, died in the classic Victorian manner. In those days the death agonies were unrelieved by the merciful injection of morphine or other pain-killing and soporific drugs. Indeed, religious folk were expected to face their end in full consciousness, expressing certainty in their faith and to be despatched by loving relatives in a vigil where scripture texts were recited and even hymns sung at the bedside. For years Mrs. Daniell had been ignoring deteriorating health to labour on at her institute. Then doctors told her that the agonising pains that racked her sprang from cancer of the breast. Nevertheless she continued her Aldershot work, concealing her sufferings behind a smiling countenance. Then, after nearly three years of defiance of the cancer she became too weak to carry on any longer.

She moved to her family home, Eastwick House, Great Malvern, Worcestershire, prepared for a devout leave-taking from this wicked world. Lying in acute pain, she sent day-by-day messages to all in Aldershot and elsewhere who were missing her. 'Tell all the dear ones', ran one message, 'that my Father is taking me home by a rough way; but I stay myself constantly on the verse, "He led them forth by the right way".' 'Do you feel Jesus near?' Georgina asked her dying mother. 'Oh yes!' was the reply. 'He is very near, though I cannot see Him with the eye of sense, yet I know He is here, and I have not a fear or doubt. I have the royal word of a king to go on.' As death drew near her, Georgina inquired, 'Isn't it sweet to be going home?' The mother answered, 'Yes, it's lovely.' And so on, with *demande et réponse* between the living and the dying until Louisa Daniell stopped breathing on Saturday evening September 16, 1871, while struggling to repeat after her daughter the verse, 'I will never leave thee nor forsake thee.'

They brought the body back to Aldershot for lying in state in the institute hall and for burial on a sunny slope of the military cemetery. Customers came out of the public houses and girls from the bawdy houses to stand in solemn respect as the funeral cortège passed. Marching beside the hearse was an escort of army engineers. There followed several carriages including that of General Sir James Hope Grant, Aldershot

G.O.C. and one of England's most renowned soldiers. Next
there were officers, N.C.O.s and privates from several regiments marching six abreast in their colourful uniforms. Lastly,
out of step, was a host of women, shabby and sobbing. The
women were present and past members of Mrs. Daniell's
mothers' meeting and sewing class. In the following month a
plain stone was raised above the grave. This was given by
officers of The Camp.

Georgina continued her mother's work for the welfare of
soldiers and their families until her own death in 1894. And
during those years she greatly enlarged the Aldershot institute
and also established branches for the troops at Colchester,
Plymouth, Chatham, Windsor, London and Okehampton.
Meanwhile, inspired by the pioneer example of the Daniells,
the Wesleyans opened a Soldiers' Home in Aldershot. This was
followed by a similar establishment managed by the Church of
England. Soon others were started by various religious and
welfare organisations. Thus Louisa Daniell was in a sense the
Florence Nightingale of the soldier's moral and social welfare,
carrying, instead of a lamp, a cup of tea in one hand and a
bible in the other. While the example of Florence Nightingale
resulted in the spread of military hospitals, the example of the
Daniells roused the country to the realisation of the need of the
troops for social and recreational amenities.

3

Royal Flush

ALDERSHOT today has a dull, dreary look. The inhabitants of the ancient town of Farnham are frequently heard to say of adjoining Aldershot, 'I would never live *there*.' Farnham, with its graceful Georgian architecture and castle, is steeped in tradition and looks it. On the other hand Aldershot, a mixture of second-rate Victorian and modern architecture, appears drab and common. From time to time councillors of the borough of Aldershot suggest that the urban district of Farnham should be absorbed into its boundaries. These suggestions are met by roars of indignation from residents of Farnham. They point out that whereas Farnham has centuries of written history and tradition, Aldershot has no history prior to the middle of the nineteenth century. As a matter of fact there is passing mention of Aldershot in the A.D. 885 will of King Alfred, but after that it was apparently forgotten like his cakes. However, even the people of Aldershot are inclined to forget that until comparatively recently their town was from 1854 onwards the glamorous magnet of the world's royalty; that Queen Victoria, Edward VII and George V all stayed often at their Aldershot residence, the Royal Pavilion. Frequently they brought foreign relatives to bask with them in the military enchantment of The Camp and to be cheered through the streets of the town. In those days Farnham seemed pitifully insignificant compared with Aldershot.

Queen Victoria, of course, regarded the army as her personal possession and Aldershot almost as her private preserve where she could mingle with her soldiers and sentimentalise over their simple loyalty. She was aided and abetted in this attitude by her first cousin George William Frederick Charles, second Duke of Cambridge, commander-in-chief of the British army for thirty-nine years. It was his influence that helped to keep the army 'in the family' despite strenuous objections and interferences from successive Ministers for War. It was the Duke of Cambridge who helped to ensure that Aldershot was made a place of splendid martial pilgrimage by the monarchs and princes of Europe, nearly all of whom were his relatives. The duke could, indeed, have claimed to be almost a foreign royalty himself. He was the son of the seventh son of prolific George III, but was born in the guttural and German environment of Hanover. Never losing a streak of Teutonism, he was frequently referred to by his cruel critic General Lord Wolseley as 'the great German sausage'. The duke commanded the army from 1856 to 1895. Biographies say he was affectionately known as 'the soldier's friend'. Nevertheless, with remarkable consistency, he obstructed army reform for half a century.

George (William Frederick Charles), the boy who rose to such dizzy heights in the British army, entered the world in 1819 when his father, the first duke, was Governor-General of Hanover. At the age of nine rosy-cheeked little George was gazetted colonel in the crack Jäger battalion of the Hanoverian Guards. Two years later he was escorted to England where, under the kindly eye of William IV, he was tutored by John Ryle Wood, later a canon of Worcester. The lad returned to Hanover for more soldiering, but this time as a mere private. When Victoria ascended the throne in 1837 George and his father moved to England. There George entered whole-heartedly into soldiering, serving at home and overseas. The father died in 1850 and George became the new Duke of Cambridge with a reluctant parliamentary grant of £12,000 a year and with growing admiration from the queen.

The duke was in many ways a human and lovable person despite a rough, gruff manner and use of bad language while in the company of fellow soldiers. Commanding a division in the Crimean War he treated the common privates under him with understanding and compassion. He was himself heroic in battle, yet it pained him to observe the sufferings of others around him. He confessed after the Battle of Alma that 'when it was all over I could not help crying like a child'. It was the same after Inkerman when overpowered by the sights and sounds of human agony, 'I felt perfectly broken down'. Moreover the Duke of Cambridge was so human that he defied the propriety of the royal family and the rules of decorum of the army by marrying an actress. In the Victorian view, being an actress was an unseemly occupation, not far in the social scale above the ignoble profession of prostitution. However good her morality and however successful at her art, an actress had to be content to suffer herself to be regarded as a female person, but never as a lady fit for the marriage bed of an officer and gentleman. Army officers defied the code of their class if they married actresses. This idea lingered far into the reign of George V when much publicity was given to the case of a Guards officer who was requested to resign his commission following his marriage to an actress.

The actress who won the heart and long devotion of the Duke of Cambridge was Miss Louisa Farebrother. He defied the prohibitions of the Royal Marriage Act to wed her morganatically in the New Year of 1840 when he was twenty and she was twenty-four. Louisa never ventured into the public life of the duke. She was for years anathema to most members of the royal family until her virtues won grudging respect. Calling herself Mrs. Fitz-George, Louisa was maintained in a house in Mayfair and she bore her husband two sons. The diary of the duke and the letters exchanged between him and Louisa revealed passionate affection and a mutual sense of romance that endured until her death in 1890. The duke would delight in his many fleeting escapes from the army and national life for days—and sometimes mere hours—with his 'secret' wife. The letters of each expressed their mutual delight in their transports of physical love, references to which were disguised by the euphemism of 'having a cuddle'. But, alas, the duke's adoration for his wife did not entirely preclude his exercise of the Hanoverian prerogative of 'having a cuddle' with other women.

Naturally such human traits in the character of the Duke of Cambridge endeared him to the Prince of Wales, the future King Edward VII. The prince affectionately addressed the duke as 'Uncle George' and refused to share the prissy disapproval of his parents and other court personages for ex-actress Louisa Farebrother, otherwise Mrs. Fitz-George. Since both Queen Victoria and her Prince Consort trusted the duke (in spite of that marriage) they relied upon him to aid in keeping Edward 'on the right path'. Often when they had advice and admonitions for Edward they would convey them via the mouth of his Uncle George. Edward would patiently listen to these, then dismiss them with a grunt—or a chuckle.

It was through the counsel of the duke that, in 1861, Edward was sent to Curragh Camp in Ireland for military training instead of to Aldershot, so close to the temptations of London. The prince went to Curragh as a colonel, but to the chagrin of the commanding officer Colonel Percy, he proved hopeless in elementary drill. Even when the camp was visited by the prince's parents and by the Duke of Cambridge, Percy allowed him only to undertake the duties of a subaltern in the welcoming parade. Then, soon afterwards, Victoria and Albert appealed to the duke to strive to save their son's morals. Brother officers of Edward at Curragh arranged for saucy and audacious actress Nellie Clifden to be smuggled one night into his rooms. The secret was badly kept and soon, in an inquiry, Edward was refusing to name those officers who had managed this immoral plot. The Duke of Cambridge, as Uncle George and also as commander-in-chief had, of course, to warn the offender against ever again indulging in such scandalous conduct. In that same year the Prince Consort died. Victoria attributed her husband's fatal illness to the grief caused him by the sexual demeanour of their eldest son.

The Duke of Cambridge played a dominating part in the expansion of the Aldershot camp until it became the talk of the militaristic world of the time. Everyone wanted to see it, from proud European monarchs and the spies they employed, to humble potentates in distant and almost inaccessible areas of Queen Victoria's expanding empire. The Camp was a panorama of colour, full of noise and excitement. In 1858 an adjective-weighted book called *Sketches of The Camp at Aldershot* described it thus:

'That great scene of animation and activity, that stupendous practical military school, with its multitudes of men surrounded by all the paraphernalia of war; its Royal Pavilion, its churches and schools, its theatres and clubs, its thousand huts, its commissariat, its daily routine of discipline and its grand field days when its troops engage in mimic warfare—frequently in the presence of royalty—when its valleys resound, and its hills echo with the sombre booming of ordnance, the sharp and incessant popping of musketry, and ominous clang of bayonet and sword, the tramp of cavalry, the measured tread of infantry, the hoarse sounding trumpets, the mellow bugle, the romantic pibroch, the shrill fife and spirit stirring drum, when may be seen the sturdy Guardsman, the kilted Highlanders, the dusky Riflemen, the powder-stained Artilleryman, the dashing Cavalryman and the fiery charger impatiently champing his bit; the bright accoutrements, the polished helmet and the gleaming sabre glittering and flashing in the resplendent sunbeam. While at intervals may be discerned floating proudly in the breeze, old England's time-honoured banners, torn by many a foeman's shot in America, the Peninsula, China, India or the Crimea—banners which have been planted as the symbols of victory on many a hard fought and bloody field, and on the walls of many a stone-girt city deemed impregnable—banners which have serenely flapped over thousands of England's brave and dauntless sons, whose hearts have ceased to beat, and who have found their resting place in a stranger's land.'

The great came to gaze and wonder at The Camp throughout the years of the Duke of Cambridge's long term as commander-in-chief, and for many years after that. The foreign royalty who caused the biggest Aldershot flutter of the 1870s was the progressive Alexander II of Russia. Alexander became czar in the middle of the Crimean War when the Anglo-French siege of Sebastopol was in its fifth month. The ex-enemy came to England on a state visit in 1874, the year of the marriage of his only daughter the Grand Duchess Marie Alexandrovna to Queen Victoria's second son Alfred, Duke of Edinburgh. This was an alliance which disturbed Prussia's Bismarck, who hated to see the British lion becoming too friendly with the Russian bear. In 1861 this Alexander had emancipated the serfs. He was allowing his millions of subjects to sniff a faint odour of democracy, with promises of more freedom in the future. The czar's foreign policies were beginning to convince the suspicious British that he was truly working for European peace. However, Queen Victoria was shocked that the middle-aged czar had taken to himself a young mistress, Princess Catherine Dolgoruka, to the pain of his wife, the Czarina Marie. The czar had three children by his mistress and after the death of Marie in 1880 he married her.

The state visit of the czar and czarina in May of 1874 was a reward to them for giving their daughter to the Duke of Edinburgh, and for their munificent entertainment of his brothers Edward Prince of Wales and Arthur Duke of Connaught at the time of the wedding in Moscow in the preceding January. Arthur, Victoria's third and favourite son, had been his brother's best man at a Russian Orthodox ceremony, full of colour and mystical ritual. The Prince of Wales was there to watch in the course of a European tour which included junkets in the courts of Emperor Franz Josef in Vienna and of Kaiser Wilhelm I in Berlin. In Russia the prince had been particularly gratified by a boar hunt organised on his behalf by Czar Alexander, often described as 'sensitive' and 'humane' by historians. At this hunt eighty of the beasts were slaughtered for the prince's pleasure.

The appearance of the czar and czarina in England caused near hysteria. Nearly all ordinary folk wanted to see the ruler of a country which was always painted as mysterious and inscrutable. On the other hand government and police were disturbed by the presence of the august couple because of the danger of deadly assault upon them by anarchists. The Duke of Cambridge appointed Prince Arthur Duke of Connaught in charge of A Troop of the 7th Hussars as special escort to the czar. Once in London a friendly mob, pressing down barricades and breaking police lines, came very close to the czar. Commander-in-chief the Duke of Cambridge was so disturbed that suddenly in stentorian tones he commanded Prince Arthur and his cavalry to 'Charge!' But Arthur ignored the command. He called his hussars to order and got them to edge their horses towards the mass of people at the same time politely suggesting that they fall back. There was more anxiety when cockney crowds in carnival mood reappeared at night merrily to jostle and insult the haughty people being driven to the great state banquet for the czar and czarina at Marlborough House where both Gladstone and Disraeli were presented to the visitors.

The czar came to Aldershot on May 19, 1874, following rumours and alarms that anarchists had seeped into camp and town. Many 'suspicious looking' people were challenged and interrogated by police or detectives, only to be proved harmless and friendly. The visit had also been preceded by weeks of intensive drill of 15,000 troops by The Camp's G.O.C., sixty-six-year-old General Sir James Hope Grant, a hero of the Indian Mutiny. On the big day the czar stood with the plump

Prince of Wales and the proud Duke of Cambridge on either side while the 15,000 in their colourful variety of uniforms paraded past him. Actually this deeply religious ruler did not enjoy watching soldiers. He had always found it difficult to overcome a childhood revulsion for anything associated with conflict and killing. Once, when a beautiful little tree withered after being transplanted, the boy wept. Then in his manhood Alexander remarked to a friend: 'Why should wars be inevitable? We believe in Christ—not Mars.' But apparently such sensitivity did not extend to boars, or to men who were ruthlessly slain or tortured if they opposed his rule of Holy Russia. General Hope Grant had survived the excitements of the Sikh wars, of the Mutiny and the Chinese War; yet the visit of the czar to Aldershot left him exhausted. In the following year his remains were given a big military funeral. And in 1881 Alexander II also died. A young man hurled bombs successfully at the czar who hated violence while he was driving through Moscow after a visit to a relative.

Alexander II was, we must suppose, comparatively mild for a royal despot. But his grandson, the Czarevitch Nicholas, was even milder. Nicholas visited Aldershot in July 1894 when he was twenty-six and during what was probably the happiest period of his tragic life. He had recently become engaged to Princess Alix of Hesse-Darmstadt. In those days most royal engagements were 'arranged', but this one was announced only after a hard struggle by Nicholas to overcome the objections of his parents, Czar Alexander III and the Czarina Marie, and even of the girl herself. The parents did not like Princess Alix's manner which they described as gauche. And the princess who later as czarina became the adoring dupe of the Orthodox monk Rasputin recoiled from marriage to Nicholas because this would oblige her to join the Orthodox Church. However, such objections were overcome and the czarevitch became engaged to Alix on what he described in his diary as 'a heavenly, unforgettable day'. In pursuit of the heavenly, Nicholas happily accepted an invitation of Queen Victoria to join his fiancée as a guest at Windsor Castle in June 1894.

Still in England in the middle of the following month, the young czarevitch stayed overnight in the Royal Pavilion at Aldershot with Queen Victoria. He and the queen watched a performance of a Torchlight Tattoo—then a novel attraction—in the pavilion grounds. The czarevitch wore a uniform of blue and gold. On his head was a scarlet cap. Nevertheless ordinary soldiers and civilians from Aldershot town were disappointed. True that his uniform was fine, but he looked too mild, too nervous to be the heir of the throne of All the Russias. But this Torchlight Tattoo was quite a family affair. Of course the Duke of Cambridge was there, together with Victoria's son Prince Arthur Duke of Connaught who in the previous year had been appointed General Officer Commanding at Aldershot. Indeed the idea of the Torchlight Tattoo with its spectacular marching and counter-marching and its huge massed bands was originated by Arthur. British officers presented to Nicholas at the tattoo found him charming but hesitant, with an apologetic manner which was almost abject in its humility.

The czarevitch made a similar impression when he visited the House of Commons to observe democracy debating. 'He seemed shy, uncertain, undecisive and went to his place with much awkwardness,' observed politician-journalist T. P. O'Connor. 'There was something suggestive of a lonely and perilous elevation to which he will attain, in this little scene—of all the solitude, desertion and uncertainty in the midst of millions of adoring subjects and thousands of servile courtiers.' Nevertheless the diaries of Nicholas testify to the heights of ecstatic happiness which he attained in the spring and summer of 1894. But this, with its comparative tranquillity was fleeting. In November of the same year Alexander died. Nicholas became czar.

Nicholas loved—and envied—the British royal family. He was specially happy in the company of his cousin George, later

King George V to whom he bore a striking facial resemblance. It was King George who vainly tried to arrange for the czar, his wife and family to be brought to the safety of England during the great Russian revolution. However, they were all horribly to perish, shot to death in a cellar.

If the bashful czarevitch Nicholas disappointed the people of Aldershot they were generously compensated almost a year later by the visit of Shahzada Nasrullah Khan, truly gorgeous son of the Amir of Afghanistan. He came by special train to Farnborough with the Prince of Wales and the Duke of Cambridge. At Farnborough the shahzada and company entered carriages for the drive to The Camp with mounted men of the 9th Lancers as escort. The uniform of the Afghan prince almost caused spectators to rub their eyes. His uniform was a loud red and overlaid on this were facings of gleaming gold, and over his shoulder was a blue sash. Aldershot did all possible to impress the shimmering shahzada. For the British government felt that a display of British might would so awe him that he would use all his influence to keep Afghanistan at peace with a queen and country who wielded such enormous power.

Shahzada Nasrullah Khan was greeted on the big parade ground known as Laffan's Plain by an assemblage of 17,700 troops with guns booming at his approach. A dozen bands blared music. The *Aldershot News* reported that, 'to add to the general scene the Balloon Section, Royal Engineers, became in evidence in mid-air, a balloon circling majestically above the parade,' while on the ground 'as far as the eye could see arose a forest of steel and pennons of the Lancers'. Later there was a 'gallop past' by the Royal Horse Artillery and a charge by the Heavy Brigade. The blimpish Victorian officers of The Camp had made an all-out effort to show this 'native' with what ease the almighty British raj could, if necessary, crush any trouble from the Afghans or any other wallahs in Asia. But to everybody's surprise, all the troops and their martial ironmongery

E

and the 'majestically' circling balloon seemed to make no impression upon the shahzada. Indeed, instead of showing concern, interest or elation he looked bored. During the 'gallop past' he was seen to yawn. The *Aldershot News* noted that when Nasrullah 'glanced along the gleaming mass of troops, his impressive face betrayed no thought, although the eyes of his bearded attendant officers gleamed with enthusiasm'. After the parade the Prince and Princess Arthur of Connaught welcomed the apparently unimpressed and emotionless Afghan to a bumper luncheon at their residence Government House. At the sight of the food he became a changed man. Now the shahzada became talkative and gay. He ate with gusto. It was obvious that this prince of a warrior state preferred butter to guns.

There are still men and women alive in Aldershot who in childhood gazed upon Kaiser Wilhelm II of Germany, grandson of Queen Victoria, in all the theatrical splendidness of his flashing eagle-topped helmet and fearsome high-glossed upturned moustache. His imperial majesty came twice to Aldershot. His first appearance, in just a passing call, was in 1889. The second—an occasion of magnificence—was in 1894. The 1889 invitation by Victoria and her government to Wilhelm to visit England was in the course of an attempt to mend a squabble which occurred soon after he had become kaiser. The squabble was caused by the inability of Wilhelm and his uncle Edward Prince of Wales to get on well together. It would seem that Wilhelm disliked Uncle Eddie even from early childhood when he caused a disturbance at the marriage of uncle to the Princess Alexandra. There the four-year-old Prussian, sitting in the choir, pulled a cairngorm (gem-stone) from the decorative dirk he was wearing. When grabbed by two other British uncles—Arthur and Leopold—he bit their legs.

But at the time of the funeral of Wilhelm's father Frederick III in 1888, the Prince of Wales was the alleged offender. Wilhelm complained that the prince disrespectfully treated him more as a mere nephew rather than as the German emperor. Wilhelm was also furious because his uncle was reported to have made impertinent and meddlesome inquiries about the future of Alsace-Lorraine, annexed by Germany from France in the war of 1870. Queen Victoria thought that Wilhelm's conduct towards Edward revealed 'a very unhealthy and unnatural state of mind'. However, on an appeal from her daughter the ex-Empress Frederick, the mother of Wilhelm, the queen decided to try to make bygones be bygones.

Thus in 1889 Wilhelm was made an honorary admiral of the British fleet, and he arrived in England on a state visit with an escort of twelve warships of his own. The proud new honorary admiral inspected 100 British naval units at Spithead. A welcome to Aldershot was arranged for the kaiser as the grand finale of his state visit. But tactfully it was whispered to Wilhelm that his Uncle Eddie would not accompany him there owing to severe phlebitis which prevented him from mounting a horse. Apparently abnormally dilated veins above the legs would have made contact between the royal bottom and the back of a charger too painful to endure, even in the cause of Anglo-German friendship. However, the day of the visit—August 7th—passed in an atmosphere of cordiality with Alexandra Princess of Wales deputising for her husband.

To a commentary from Aldershot G.O.C. Lieutenant General Sir Evelyn Wood and commander-in-chief the Duke of Cambridge, Wilhelm watched a mock battle which engaged 26,000 troops. This was followed by a grand parade past the visitor. The presence of a stout German admiral in the kaiser's entourage at least contributed some hilarity to the day. In a gallop towards the carriage of the Princess of Wales he halted his charger so abruptly that he was vaulted into the air, turning a full somersault before landing on the grass. Soon afterwards, towards the end of the review, the naval officer fell off his horse again when it shied before a prickly gorse bush. 'By jove,' remarked Wilhelm to Sir Evelyn Wood in his excellent English, 'there's the admiral overboard again!' On saying good-bye Wilhelm added that he hoped very much to return to Aldershot one day. And he later sent Wood a breath-taking gift from Berlin. This was a German sword bearing the imperial monogram and crown. The hilt of the sword was studded in diamonds.

The kaiser's hopes for a return to Aldershot were fulfilled in August 1894 when he remained there thirty-six hours, with almost every minute of daylight packed with activity. By that year Wilhelm had grown in importance in the eyes of England, including Aldershot. In 1889 the people of Aldershot felt they were bestowing an honour on the fledgling emperor by receiving him. But in 1894 the townsfolk felt that, by coming back, Wilhelm was honouring them. Thus a committee was appointed to supervise welcoming decorations at the railway station and throughout the town. Drapers sold roll after roll of coloured bunting, together with hundreds of miniature German flags and union jacks. One shop sold toy German army helmets made of cardboard and brushed with silver paint. Small boys, many of whom were to die fighting against the Germans in the 1914–18 war, placed these helmets on their heads and tried to look proud and imperious as Wilhelm did in his portraits.

The kaiser arrived at Aldershot by a special train where he had occupied a saloon specially upholstered in cream satin. But, to the disappointment of the townsfolk, particularly the small boys, he was dressed merely in the uniform of a colonel of Britain's Royal Dragoons. Queen Victoria had, to his great pleasure, made him colonel-in-chief of that regiment. However, with the British-uniformed Wilhelm came a big group of German officers in Teutonic ensembles that roused gasps of admiration. Their uniforms were of Mediterranean blue and ornamented with a mess of gold and silver lace. Their Pickelhaube (spiked) helmets matched the toy replicas on the heads of the small Aldershot boys. A mint of decorations and medals added jewel-shop grandeur to the chests of the German guests, and of course there were ribbons and sashes in plenty. The kaiser's ribbon was that of the Royal Order of the Garter. Prince Arthur Duke of Connaught as new Aldershot G.O.C. was on the platform to meet his nephew Wilhelm together with a personal staff which included General Sir Redvers Buller who would, in 1898, himself become the G.O.C. Prince Arthur wore the Yellow Sash of the Order of the Black Eagle with which his nephew had rewarded him. There was an orgy of stiff saluting, bowing and clicking of heels as if these grotesque puppets of the imperialist age, now long dead, were being manipulated on strings.

After inspecting a guard of honour of the 5th Northumberland Fusiliers on the bunting draped and beflagged station platform, Wilhelm moved to the entrance where awaiting him was a shimmering black charger held by a handsome young groom in blue and silver livery. Welcoming guns boomed. The kaiser, despite one withered arm, gracefully mounted the horse from which he inspected 'his regiment', the 1st Royal Dragoons. In attendance upon the kaiser on horses as fine as his own were General von Plessen, Colonel von Scholl and Major Count Moltke. And as the guests moved into the street they saw an enormous notice of gold imprinted on velvet reading: 'A Joyous Welcome.' Next they saw a large flag bearing the reassuring words 'Gott mit uns' (God With Us) quoted from the kaiser's imperial standard. At one street corner came cries of 'Hoch der Kaiser!' from members of Aldershot's German community of butchers, sausage vendors, barbers and musicians.

When the procession reached Laffan's Plain the kaiser found 12,000 troops there to meet him. Massed bands played the British and German national anthems. After a period in the saluting box Wilhelm rode down the lines of troops to get a closer look. Finally, there was the inevitable 'march past'. In the afternoon Wilhelm visited Farnborough Hill, the home of the Empress Eugénie, to pay his respects to this widow of Napoleon III who was made prisoner by the conquering Prussians in the war of 1870. And in the evening Prince Arthur gave him a banquet when toasts were drunk to eternal Anglo-German friendship. The banquet was held in Government House, the mansion built three years previously as official residence for the G.O.C.s in place of a modest wooden building they had previously occupied.

Guests at the banquet commented upon the fresh and healthy appearance of the emperor after the rigours of the long day. Perhaps he owed some of his fresh look to the care of the three valets, a hairdresser and the hairdresser's assistant who travelled with him everywhere. The duty of the hairdresser's assistant was to ensure that his imperial master's moustache remained firm, with its two ends curled upright towards the heavens. Wilhelm's bearded uncle, Edward Prince of Wales, knew all about this Keeper of the Imperial Whiskers. He made sneering and jocular remarks about this hairdresser's assistant. However, the many men who at that time wore waxed and curled moustaches could have testified that, unless constantly attended they would wilt and lose shape, particularly in wet or damp weather.

The kaiser's Aldershot visit was enlivened by the mystery of one of his gee-gaws—the Gold Chain and Order of Hohenzollern. He was wearing the order among the mass of other decorations on his arrival at the railway station but later in the day found it had vanished. When he first reported the loss there were fears it had been stolen, but only a wizard of a thief could have removed it from the kaiser's uniform without attracting attention. However, in a widespread search the missing Gold Chain and Order of Hohenzollern was found shining on the grass at Laffan's Plain where it must have fallen while their owner was inspecting the troops.

On the following morning Wilhelm, again on a horse, watched a mock battle. He lunched in the Royal Artillery mess in Waterloo Barracks. In the evening he was at an army boxing display and dining in the mess of the Scots Greys. The British army at Aldershot liked Wilhelm and was genuinely sorry when his visit ended. Officers and men—and he chatted with many of them—found his manner far more natural and more charming than his proud imperial countenance. The German officers who accompanied him were admired for their simple friendliness and for their hilarious jollity over drinks.

As for the citizens of Aldershot, they made a gala night of the kaiser's departure and gave him a farewell which he must have often recalled. He and his party, with British hosts and escorts, started for the railway station at 10 p.m. They entered a town filled by thousands of men, women and children, cheering wildly. The town had been turned into a fairyland of illuminated decorations. All along the route were strings of coloured lights. The station itself blazed with 3,000 coloured lamps, while at the entrance was a giant illuminated star of red and amber. When the party entered the station, fireworks were bursting and illuminating the skies from open spaces in all parts of the town. In the offices of the *Aldershot News* the staff worked all night on a special illustrated supplement on the kaiser in Aldershot. A few days later a copy was forwarded to him in Germany. It was printed on yellow satin.

When Queen Victoria emerged from many years of semi-seclusion which followed the death of her beloved Albert she came to Aldershot almost every year, usually staying a night or more at the Royal Pavilion. She would normally travel by train to Aldershot or Farnborough where she transferred to a carriage for the final lap of the journey. The carriage was pulled by four horses, complete with mounted postilions. There would be outriders in royal livery, together with an escort of cavalry. On these Aldershot jaunts she invariably brought one or more of her multitude of European relations. After the death of Germany's Frederick III in 1888 a frequent companion of Victoria was his widow, known as the ex-

Empress Frederick. Formerly the Princess Royal, the empress had the honour of being a daughter of Victoria as well as mother of Wilhelm II. Victoria and daughter mixed watching the troops at The Camp with homely gossip about old times and about their network of relatives.

The biggest concentration of royalty ever seen with the queen at Aldershot was during her Diamond Jubilee celebrations in 1897. These and a host of other notables from all over the world formed an almost walking edition of the *Almanac de Gotha*. To mention just a few names those in close proximity to the queen included the Prince and Princess of Wales, the Duke and Duchess of York, Princess Victoria of Wales, the ex-Empress Frederick, Prince and Princess Charles of Denmark, Prince Henry of Battenberg, the Duchess of Albany, Princess Victoria of Schleswig-Holstein, Princess Aribert of Hainault. In addition there were hosts of other princes, princelings, princesses, peers, statesmen, generals, admirals, military and naval attachés. India was represented by a row of maharajahs, Africa by an assortment of sultans and native chiefs. Lastly, gathered to watch were thousands of the queen's ordinary subjects from all over the British Isles.

Nearly 30,000 troops participated in this Diamond Jubilee military review. More than 1,000 of these came from 'the colonies', meaning Canada, Australia and South Africa. These colonials were given the honour of leading the procession. Patriotic emotion among troops and civilians reached its warmest after the royal salute of the infantry. Helmets, bonnets and busbies were waved on the points of bayonets or just tossed wildly into the air. Civilians, in emulation, threw into the air their top hats, bowlers and cloth caps. Much good headgear was sacrificed to patriotism and trampled into permanent deformity as the crowd surged like ocean waves in their cheers and cries of 'God save the queen!' God did save the queen until her death in 1901. Two years before that she paid her last visit to Aldershot. On that occasion she emerged from the train

leaning hard on the arm of a tall Indian servant. Her little body
was trembling. She walked with difficulty. She stayed less
than three hours, then tottering back on the train left Aldershot
for ever.

The Duke of Cambridge was, of course, with the queen at
Aldershot for the Diamond Jubilee review of 1897, but under
circumstances which must have caused him great mental pain.
He was no longer commander-in-chief of the army, having
been superseded in June 1895 by his clever, scheming rival
Field Marshal Lord Wolseley. The unhappy duke, after
dominating the army for thirty-nine years, suffered the galling
experience of having to go to Aldershot on this historic occasion and see his triumphant rival, who had called him by so
many rude names, strutting and smiling. The troops at this
Diamond Jubilee were under the command of Prince Arthur
Duke of Connaught, who was still The Camp's G.O.C.
Cambridge and Connaught had run these Aldershot affairs
together until after the visit of the Shahzada Nasrullah Khan
early in June 1895. Then, later that month, Cambridge was
forced to resign. And one painful afternoon all the army headquarters staff were assembled in a room at the War Office. The
seventy-six-year-old duke entered and said he had come to say
good-bye. He broke down, covering his eyes to hide his tears.

The famous old song tells us that 'old soldiers never die' but
'simply fade away'. The tragedy of the Duke of Cambridge
was in his refusal to fade away when it was obvious that as a
septuagenarian, with attitudes and ideas coagulated by time, he
was no longer in a fit condition for supreme command. Yet he
fought a long, dogged battle to maintain that command. It
required constant pressure, not only from Wolseley but also
from the Prince of Wales, from War Minister Campbell-
Bannerman, together with some firmness from the queen,
before the gallant old soldier surrendered. The war between
the Duke of Cambridge and the advocates of military progress
had been long. After the Franco-Prussian War, Lord Cardwell,

Minister for War in the Gladstone administration, battled for his famous reforms which included the abolition of the purchase of commissions, the retirement of ageing officers and the short-service system. But the commander-in-chief argued that commissions should still be sold, that elderly men should still be active officers and that there should be no short-service system. Such revolutionary changes, insisted the duke, would destroy the *esprit de corps* of the army.

However, a clever and unrelenting opponent of the Duke of Cambridge was General Sir Garnet Wolseley, raised to the peerage in 1882. As Assistant Adjutant General, Wolseley had aided and abetted Cardwell in his reforms. A fair strategist and an able organiser, Wolseley had moreover won the affection of the nation as an enterprising and daring commander on active service in Canada, Africa and Egypt. At the War Office Wolseley had gathered about him a group of young officers who in their time were regarded as 'bright and progressive'. Since military historians of today question the ability and achievements of these officers it might be said that they were 'brighter and more progressive' than their contemporaries in an age when the British officer was mediocre in strategy even if brave in battle. The group assembled by Wolseley were popularly known as 'the Garnet ring'. Wolseley and his 'Garnet ring' treated their commander-in-chief with humorous disdain. In addition to referring to the duke as 'the great German Sausage', Wolseley wrote of him as 'the dear old Bumble Bee' and that 'Mountain of Royalty'. Wolseley complained that the commander-in-chief had to be 'coaxed and flattered and terrified as one would act in dealing with some naughty little girl or some foolish old woman'.

Forced to resign at the age of seventy-six, George Duke of Cambridge made way for new-fangled ideas of organisational progress and efficiency. Yet with the old man there departed from the army a spectacular grandeur and panache so exemplified in the imposing military reviews before richly uniformed

monarchs and their military attendants. The duke liked to play
with live soldiers in the same way as a boy will play with his
painted metal collection on the floor of the sitting-room. His
failure was that in his passion for reviews and parades, fancy
drilling and marching, he forgot that the real function of the
troops was to learn to fight and not merely to show off.
Throughout the many years of the duke's command British
military strategy, tactics and organisation were kept marking
time. That is why some of today's best military experts point
accusing fingers at the old duke and hold him largely responsible for the backward and mediocre leadership of the South
African War and then, years later, of the First World War. He
had kept the army marking time for so long that it could not
catch up with progress.

For most of his years as commander-in-chief the duke was
idolised by many officers and men, who among themselves
affectionately referred to him as 'old Pork Chops'. Once on a
visit to a regimental mess the duke praised the pork chops with
which he was served. The news spread and thereafter wherever
he was entertained pork chops were invariably on the menu.
Another quality in the duke that endeared him to his men was
his ability to swear like a trooper. Once complaints of foul
language used by all ranks reached Queen Victoria and she
spoke to the commander-in-chief about it. Next morning the
duke, addressing a big parade, dutifully but fiercely commanded that cursing and swearing be curbed forthwith. He
added: 'I was talking it over with the queen last night, and her
majesty says she is damned if she will have it.'

Lord Wolseley was full of ideas for new changes and new
reforms when he succeeded 'the great German Sausage' as
commander-in-chief, but to his chagrin he found that the
powers attached to that office were being drastically reduced
by the government. Wolseley was sufficiently tactful not to
try to interfere with Prince Arthur Duke of Connaught, who
remained at Aldershot as G.O.C. until 1898 when he was

succeeded by the renowned General Sir Redvers Buller. Prince Arthur was as a personality unspectacular. Throughout his long service to the British army and empire he endured a reputation of being awfully good but terribly dull. However, there can be but one opinion of his work at Aldershot. It was competent, and he was beloved. It seemed that in administering The Camp he bore in mind the advice given to him by his doting mother Victoria in 1871. She wrote: 'Continue to be kind and considerate to those below you, and treat those who faithfully serve you as friends. . . . It is by those below us that we are most judged, and it is of great value to be beloved.'

From boyhood Arthur loved soldiering. Queen Victoria would watch the child she called her 'ray of sunshine' playing in the miniature fortifications erected for his pleasure at Osborne. Growing up, he made the army his profession and practised soldiering with earnest fervour. His first appearance at Aldershot was in 1873 at the age of twenty-three. There he was attached to the staff of an infantry brigade. During manœuvres he slept in a hut of which he wrote: 'It was so small I could lie in bed and open the window and poke the fire.' In 1879 he married Princess Louise Marguerite, youngest daughter of Prince Frederick Charles of Prussia. He and his bride made a ceremonial entry into Aldershot to live at the Queen's Pavilion while their permanent home, Bagshot Park, was being prepared for their occupation. He was then colonel-in-chief of the Rifle Brigade. The town of Aldershot presented Louise with an illuminated address. This told her she was not only welcomed 'as the bride of a son of our well-beloved queen, but as the bride of a soldier who has endeared himself to the whole army by the zealous way he has devoted himself to the duties of the honourable profession'.

Anything but a chocolate soldier, Prince Arthur regarded living dangerously as one of his bounden duties. In the 1882 expedition to Egypt he was mentioned three times in despatches for his gallant leadership of the Guards Brigade at the battle of Tel-el-Kebir. Later he became commander-in-chief in Bombay and then took over the southern military district of England until in 1883 he was appointed Aldershot G.O.C. The prince and his wife strengthened their reputation with the army for being dull but at the same time strangely lovable. He was totally devoid of any official arrogance or pettiness, while she could talk as easily with a private's wife as with the wife of a general. There are today in The Camp many unofficial memorials to Prince Arthur in the form of rows of chestnut and other trees. When the queen complained that her troops were without shade on the roads and parade ground of The Camp her son started a big tree-planting programme. The princess was specially interested in hospital welfare. Aldershot's Louise Margaret Hospital was named after her.

On one occasion Aldershot's zeal in paying honour to the popular wife of the G.O.C. ended in weird disaster. It had been decided that a large new balloon constructed by the Balloon Section of the Royal Engineers should be named the 'Duchess of Connaught'. The duchess was invited into the balloon yard to perform the christening ceremony. The afternoon was hot and sultry when Louise, accompanied by Arthur, stood in a decorated shed before advancing to the gleaming new balloon. Suddenly there arose a big thunderstorm, and lightning struck the balloon. Current coursed down metal ropes of the balloon to the hands of the three men who were gripping them. The three were badly burned. While thunder still roared and lightning flashed, Prince Arthur rushed to the spot and supervised the examination and removal of the injured men in the pouring rain. The duchess was terribly upset, and daily, in the ensuing weeks, she would inquire how the burned men were progressing and send them messages.

Aldershot, when Prince Arthur was its G.O.C., was tremendously social. This may have been due to the hunting and social activities of his predecessor, General Sir Evelyn Wood, who will be studied in a later chapter. Wealthy officers and their wives rented large country houses in the district for 'rest' and entertaining. The neighbourhood was a paradise for hunting-officers and their womenfolk. In a typical week of February 1894 the following hunts were listed on the notice board of the Officers' Club in The Camp: Her Majesty's Stag Hounds, Chiddingfold Hounds, Mr. Garth's Fox Hounds, The 'H.H.' Fox Hounds, the Vine Hounds, the Aldershot Divisional Foot Beagles. Grand balls and house parties were numerous. At the same time officers and gentlemen complained of the rising cost of living and of the gross conduct of the Chancellor of the Exchequer, Sir William Harcourt, in raising income tax from sevenpence to eightpence in the pound. Still, an officer with five shillings in his pocket could go to an Aldershot restaurant and spend it on what was advertised as 'A Dainty Dinner'. This consisted of:

> Barley cream soup
> Broiled haddock
> Scotch collops
> Braised goose with vegetables
> Potato salad
> Plum pudding
> Cheese pastry.

Even if a private soldier had possessed sufficient money to eat this meal he would not have been admitted to the restaurant where it was served. For N.C.O.s and privates were barred from the superior hotels and eating places of both Aldershot and Farnham. At the then exclusive Bush Hotel in Farnham privates bearing messages for officers who were inside were refused admission to the lobby. A General Smyth, protesting against such ruthless discrimination, wrote to a newspaper about 'the son of a titled lady' who on failing his examination for Sandhurst enlisted as a trooper in a cavalry regiment. His visiting mother asked for a private room in an Aldershot hotel in order to entertain him. On learning that this young man was not an officer, the hotel manager refused the room.

In his army career Prince Arthur Duke of Connaught had one big ambition. This was to become commander-in-chief of the British army. Certainly as a dedicated soldier, with considerable ability and the gift of leadership, he was hardly unfit for the post. However, outside the army, the nation was almost imperceptibly becoming more democratic, with a small but chilly wind of republicanism cooling the warmth felt for the queen. It was felt that to place the army in the hands of Victoria's favourite son, although so dutiful and virtuous, would be unwise and lead to national dissension. Nevertheless he was in 1900 made commander-in-chief of the army in Ireland, with promotion to field marshal in 1902. Later he became Inspector-General of the forces, and then commander-in-chief in the Mediterranean. In 1911 Prince Arthur was appointed by his nephew King George V as Governor-General of the dominion of Canada where he served with dullness illuminated by self-effacing tact. Queen Victoria once commented that Arthur, her 'ray of sunshine', has 'never given us a day's sorrow or trouble'. Arthur's motherland, with Aldershot in particular, could have sincerely said the same of him.

4

Prince Imperial

ALL Aldershot was excited in the spring of 1875 by the news that the 24th Brigade Royal Horse Artillery was to have a new, very special subaltern named Napoleon Louis Jean Joseph Bonaparte, better and more briefly known as the Prince Imperial. The candidate for the brigade was, moreover, being recommended and sponsored by Queen Victoria and the Duke of Cambridge. The commanding officer of the 24th was Lieutenant-Colonel Whinyates, who privately did not feel pleased and honoured by the advent of the Prince Imperial, although, of course, he expressed his 'joy' about it to his queen and to the commander-in-chief. Whinyates was only too well aware that the queen adored this short, dark nineteen-year-old son of the late and ill-fated Napoleon III of France. Thus under the queen's protection the youth would be free to misbehave with impunity and perhaps undermine the discipline of the brigade. It was pointed out to the colonel that the prince's record as a cadet at the Royal Military Academy in Woolwich had been good, except for wild capers when off duty. Still, Whinyates remained worried that the prince should, as it were, have been foisted on his brigade.

The Prince Imperial duly presented himself at Aldershot, reporting to Colonel Whinyates with precision and respect. The colonel summoned Major Ward Ashton, commander of E Battery, and ordered him to take charge of the new subaltern. In turn Ward Ashton sent for twenty-six-year-old Lieutenant Arthur Bigge, one of the battery's most trusted and promising officers. 'Here's our new man,' he said of the Prince Imperial.

'Now bring him into line.' No time was lost on ceremony. Lieutenant Bigge started the new subaltern on a hard and intensive course of instruction. The pupil showed great keenness and obvious ability. Within a few days friendship developed between the prince and Bigge. This friendship eventually had extraordinary results in the career of Bigge, who was one of twelve children of an obscure country parson.

Before proceeding further with the story of the Prince Imperial's life as an officer at Aldershot we shall survey the eventful and dramatic years of his life which preceded this. In Chapter 1 was mentioned the suspicions aroused for a time in Victoria and Albert by Louis Napoleon Bonaparte when, after seizing power in France, he made himself the Emperor Napoleon III. The new monarch, at that time so unpredictable, was a son of Louis Bonaparte, made King of Holland by his brother the great Napoleon, who died as England's captive on the island of St. Helena. His mother was Hortense, daughter of the Empress Josephine. No wonder that Victoria and Albert, until effectively reassured, had feared danger from the great Napoleon's namesake. In 1853, when secure on the throne of France, this Napoleon III married the beautiful Eugénie, daughter of the Spanish Count of Montijo. Three years later Eugénie, who through her mother had some Scottish blood in her veins, gave birth to the Prince Imperial. The date of the birth—March 16, 1856—incidentally preceded by fourteen days the signing of the Treaty of Paris. This treaty formally ended the Crimean War in which French and British troops had fought side by side against Russia.

The life of the Prince Imperial began—as it also ended—in tense drama. The Empress Eugénie's labour pains began on a Friday and she was in continuous agony until her son was born on the following Sunday. In her long screaming hours of labour the doctors were in a panic and at one time doubted whether both mother and baby could live. Chief palace doctor Darralde confronted the emperor to inquire, if a choice had to

F

be made, whether he should save mother or child. 'Save the child,' replied Napoleon, who, according to eyewitness the Earl of Malmesbury, 'sobbed and cried without ceasing for fifteen hours'. An intimate hour-by-hour account of Eugénie's sufferings was prepared for Queen Victoria, who, a child-bearing expert herself, always relished the fullest and most intimate details of difficult confinements. It was at three o'clock on the Sunday morning when the agony of the empress ended in the birth of a blue-eyed boy, who, testified a now laughing Dr. Darralde, was strong and very healthy. Eugénie lay behind a curtained bed in exhaustion. Napoleon went almost mad with joy. He ran about the palace hugging every person he met, from aristocrats to housemaids. He ordered 100,000 francs to be distributed among the poor. The guns of Paris fired 100 volleys. In England, as soon as Queen Victoria was awake, she was informed that a baby son had at last been borne to Eugénie. And that day the queen wrote in her journal: 'Excellent news of the dear Empress, but distressing ones of her confinement which must have been an awful one.'

The first years of the Prince Imperial's life could be described as awesome if not awful. His royal parents played with him as children would with a doll, dressing him up in all kinds of miniature military costumes. They nicknamed their living baby doll 'Loulou', an infantile variation of his third Christian name Louis. Soon after birth Louis appeared before a gathering of the nation's senators and deputies with the sash of the Legion of Honour across his long garments and a cross on his stomach. 'I name him Child of France,' Napoleon told the doting parliamentarians. 'I call him this in order to teach him what his duties are.' While still being suckled, the Child of France was given a commission in the 1st Regiment of the Imperial Guards. He was but nine months old when he appeared at military reviews, in a carriage with his mother, in the uniform of a guards officer, complete with bearskin almost hiding his little face and head. At the age of three he appeared strapped in front of his father the emperor on a charger, reviewing the returned troops who had triumphed in the Italian campaign of the Franco-Austrian War. The uniformed infant removed his ribbon of the Legion of Honour and threw it at an heroic regiment of Zouaves while everybody cheered.

At eight Louis was writing to congratulate his father on the success of the Mexican campaign. 'I am delighted at it,' emphasised the small boy. 'Yesterday I told the soldiers in the guardroom of Mexico's surrender, and everybody was delighted.' And added to this letter was the postscript: 'This morning my horse had a kicking fit, but I sat tight and did not fall.' When Louis was eleven he had his own military staff, consisting of one general, three colonels and a naval captain. He had also a pleasant English nurse named Miss Shaw and, from the age of three, a tutor named Augustin Filon.

Napoleon III tried strenuously to ape the military dedication and prowess of his great-uncle Napoleon I. It was pathetic that he and Eugénie should have contrived to make a military protégé of their little boy. The couple shared an obsession to convince the French people that Louis, as the genuine 'Child of France', gloried in war and, of course, in the martial memory of his great-uncle Napoleon I. It would seem that his parents contrived with Nurse Shaw, tutor Filon and members of the boy's military entourage to invent or at least to embroider piously heroic stories about him to be disseminated among the French public. There was the story about the empress and Louis being in mortal danger when their yacht *Chamois* ran aground off St. Jean de Luz. As waves broke over them in the darkness (said the story), Eugénie hugged the boy to her bosom saying, 'Loulou, don't be afraid!' Calmly the boy replied: 'A Napoleon is never afraid.' Another anecdote, told by Miss Shaw, was that one night she entered the bedroom of Louis to find the bed unoccupied. She crept into the large adjoining room where she saw Louis kneeling before a glass case where Napoleonic relics were displayed. In his hand was the sword worn by his great-uncle in the Battle of Austerlitz. The boy was praying, and Miss Shaw reported that she heard him say: 'Dear God—let his memory protect me.'

Such heroic anecdotes concerning the boy prince may reasonably be doubted. It is certain, however, that he shared the sense of fun and mischief of his mother Eugénie. The empress could, indeed, at times be amusingly unregal. Once in a palace corridor she slapped the face of one of the sentries, trained to stand at attention and absolutely motionless. She wanted to ascertain whether such a slap would shock him into moving. He did not move and his expression remained unchanged. When handed a reward of 500 francs for his imperturbability, the sentry waved it away with the comment: 'It is honour enough that the hand of the empress should have rested on my face.' On her return from a hunting party at Rambouillet, Eugénie slipped away to the kitchen and stuffed her pockets with flour. Again mingling with her guests, she suddenly showered them with handfuls of the flour.

Louis was healthy, happy, and allowed to mix naturally with a boisterous group of boys and girls who were mostly children of members of the royal entourage. But the prince's closest companion was Louis Conneau, son of Dr. Conneau, long a beloved friend of the emperor. Conneau was one of a little group who were taught with Louis in the palace school room. This also included Marie and Louise, the two daughters of Eugénie's deceased sister, the Duchess of Alba. Known as Marquinita and Chiquita, these sisters were delightful but nevertheless infinitely mischievous. Louis and his friends would romp wildly all over the palace and play practical jokes on stuffy courtiers and dignified ladies-in-waiting.

Louis was aged but fourteen when in 1870, as the toy soldier of his parents, he had to put on one of his numerous uniforms and leave for a real war. The was the Franco-Prussian War, which so swiftly and pitifully ended the meteoric career of Napoleon as monarch and conqueror. The war originated from the anger of a jealous France at the gift of the Spanish throne to Leopold, a Hohenzollern prince. Because of French protests the German Leopold refused the throne. Then Napoleon foolishly demanded a guarantee from the King of Prussia that no Hohenzollern should ever become ruler of Spain. The demand was refused, Prussia complained of being deeply insulted and the terrible and unnecessary war followed. The military experts of most nations believed Napoleon would be an easy victor. But although Eugénie was optimistic of a quick triumph by the glorious French army, Napoleon was privately depressed and uncertain of the war's outcome. His misgivings were made all the more painful by the existence of a torturing stone in his bladder.

As for the Prince Imperial, the war made him joyful. He was weary of the schoolroom and craved greater thrills than romps and escapades with Louis Conneau, Marquinita and Chiquita. And horrified by the prince's pastime of leaping from window-sill to window-sill high up in the palace at Fontainebleau, his tutor Filon had complained that 'his greatest joy is to court danger deliberately and in cold blood'. The first stage of Louis's departure for war resembled that of a Victorian English boy leaving for boarding school. His mother personally packed his tin trunk in his room in the Fontainebleau palace. She also took the reins in the pony-trap in which the prince and his father travelled to the tiny station at St. Cloud. The boy was smiling gaily when, as they boarded the train, the empress bid him good-bye. She clasped Louis to her bosom and then, with her index finger, made the sign of the cross on his forehead with the admonition, 'Louis, do your duty.'

The Prince Imperial's war began with a preposterous French engagement with the Germans near Saarbrücken just across the frontier. It has been alleged that Napoleon, as commander-in-chief of the French army, arranged this foray specially to give his son an early baptism of fire. They watched from their glossily painted carriage as 60,000 men on foot and horseback charged up a lightly defended ridge. Louis in his enthusiasm waved them on with his peaked army cap as if rallying on his friends at a school sports day. Within a short period the French troops had retired from the ridge back to Metz, the point of their departure. However, the occasion and given the boy a genuine baptism of fire. Bullets had fallen near him. He had picked up one to keep as a souvenir. 'His coolness was admirable,' Napoleon wrote to Eugénie who back in Paris was acting as his regent. 'He showed no emotion, and might have been strolling in the Bois de Boulogne. There were men who wept at the sight of his calmness.'

Soon the armies of France had more to weep about than the calmness under fire of a fourteen-year-old lad. Disaster after disaster pursued and overwhelmed the troops of the ailing and bungling commander-in-chief Napoleon III. The boy Louis, who had first enjoyed the war as other boys enjoy a fight with tin soldiers, became disillusioned and angry. After Weissenburg, the first big engagement of the war, Napoleon was telegraphing his wife the regent: 'Our troops are in full retreat. Nothing remains but the defence of the capital.' But Eugénie insisted upon her beloved Loulou remaining in the area of battle. When the boy was spending the night in a villa at Longueville his father awakened him in the morning with the command: 'Get up, Louis, we must go. The Prussians are shelling the villa.' The boy gazed from his bedroom window at three officers who were calmly enjoying their morning *café complet*. A shell exploded and enveloped them in smoke. When the smoke cleared Louis saw three charred corpses littered with the smashed table and cups. Soon father and son in their ornate royal carriage were being jeered by the troops in bewildered movements behind the lines. At last father and son parted. Napoleon rode brazenly but sadly for five hours amid the Battle of Sedan, praying that he might be killed. And then came his letter of surrender to the triumphant King of Prussia. 'Sire, my brother,' he wrote. 'Having been unable to die among my troops, there remains nothing for me to do but surrender my sword to your Majesty.' And when Louis was told his father was a prisoner he covered his face with his hands muttering, 'All is lost—even honour. Let me die.'

But the officers around the Prince Imperial did not risk his life any longer. They removed his army uniform and replaced it with a shoddy ready-made suit. They smuggled him in an old bus from Mauberge to the Belgian frontier and then in a train to Mons, third class. A message from the captive emperor ordered that his son be taken at once to England. It was on September 6, 1870, that early in the morning Louis was taken from his Ostend hotel to board the cross-Channel steamer *Comte de Flandres*. Soon after midday he landed at Dover where he was taken by train to Hastings and whisked into a room at the Marine Hotel. Meanwhile, with wild riots raging in Paris and cries of 'Death to that Spanish whore!' Eugénie sought the aid of Dr. Evans, an American dentist, who smuggled her in his brougham to an hotel at Deauville. Then the dentist arranged with Sir John and Lady Burgoyne, whose yacht was in Trouville harbour, secretly to take the empress to England. She boarded their yacht *Gazelle* at midnight and sailed through a terrific storm to Ryde, Isle of Wight. It was late on the night of September 8 that Eugénie rushed into the Marine Hotel at Hastings where she clasped Loulou to her breast, laughing and crying hysterically.

The empress and her son settled at Camden Place, Chislehurst, Kent, leased to them by a wealthy and ardent francophile named Nathaniel Strode. Napoleon was still held prisoner by the Prussians. Nevertheless Eugénie lost no time in forming a little court in exile with the help of those friends and courtiers who had succeeded in escaping from a France now torn by revolution as well as by war. Filon, the tutor, was in Chislehurst. The Prince Imperial's best friend Louis Conneau came too, together with his best young female friends, the Spanish Marquinita and Chiquita. Lessons and the fun and the practical jokes were resumed. But there were many interruptions because of the stream of important visitors now flowing unceasingly into Camden Place. The Prince and Princess of Wales (later King Edward VII and Queen Alexandra) came. In the following month there appeared Queen Victoria herself. She met Louis for the first time, later writing in her journal: 'The Prince Imperial is a nice boy but rather short and stumpy.' Next Eugénie and Louis visited the queen at Windsor Castle. The empress sobbed, and Victoria recorded: 'I pressed her poor little hand and took her to the audience room.'

The Franco-Prussian War ended. France, defeated and humiliated, was a republic again. Napoleon had written to his wife: 'It is in England that I wish to live with you and Louis— in a little cottage with a bow window.' On March 20, 1871, Napoleon arrived in England, released from captivity. The man who had asked for seclusion in a cottage was given a huge reception by thousands of Britons when he landed at Dover. *The Times* thought their unphlegmatic behaviour 'deplorable'. When Napoleon took Eugénie in his arms the mob cheered with delight. When he kissed a calm, smiling son women sobbed and cried, 'How beautiful!' Once settled in the luxury and spaciousness of Camden Place, with old servants and courtiers around him, Napoleon recovered his self-confidence and strength of personality. Many believed that the time would come when he would regain the throne of France.

Louis, however, became a nagging worry to his father. In his studies he was bored and listless. His standard of scholarship was low. Faced with a mathematical problem he would stare into space in bewilderment, disturbing his tutor Filon with the confession that he simply did not know what it was all about. But as soon as Louis escaped Filon and his school books into the society of other people, he seemed to become another person. There had evolved in the youth an all-conquering charm. He was smaller than average in height. He had a pleasant face but no startling good looks. But his nature, even at its most

mischievous, possessed a beauty and attractiveness that drew
love and respect from his friends and acquaintances, from
Queen Victoria and the Prince of Wales to the lowliest servant
at Camden Place. English Roman Catholic priests attributed the
secret of all this charm to a spiritual quality hard to define. It
was not until after the death of Louis that even his mother
became aware of the deeply religious and mystical character of
her son.

It was a mutual friend of Napoleon and the Prince of Wales
who suggested that Louis should apply to become a gentleman
cadet at the Royal Military Academy at Woolwich, then the
great centre of training in engineering and gunnery. When the
academy—generally known as 'The Shop'—was mentioned he
reacted with enthusiasm. His listlessness, and what Filon
thought was lack of understanding, vanished. He worked with
keen concentration for the entrance exam to The Shop—and
he passed—together with his friend Louis Conneau. As a cadet
at the academy the prince was an unqualified success. At drill,
at examinations, parades and inspections he shone. Major-
General Sir Linton Simmons, governor of the academy,
marvelled that a youth with such a strange and un-British
background should become so easily and painlessly integrated
with his British cadets. Louis differed from the others only in
that he did not play games.

Louis refused to be hedged by royal dignity. He never
protested when a band of senior cadets tossed him fully dressed
into a cold bath. With other juniors, or 'snookers' as they were
called, he ran the gauntlet of cane-swinging seniors and was
driven in undignified steeplechases over tables and chairs. He
loved to participate in the absurd pranks and practical jokes so
beloved by English students. Once, during a lecture, he climbed
up a pillar to a beam near the ceiling of the hall and roosted
there like a cockerel until angrily ordered to descend. He joined
the academy's Alpine Club, members of which would creep
out at night to ascend to the roof-tops and hang chamber-pots

there in artistic clusters. Indeed, he became the academy's best expert in the difficult art of 'chamber arrangement'.

Louis was attending a mathematics lecture at the academy on January 9, 1873, when an emissary from Camden Place entered the room and told him that his father the emperor was dead. Napoleon had undergone two bladder operations and had suddenly collapsed while awaiting the third. There was a huge funeral from Camden Place, attended by hundreds of French Bonapartists and by the Prince of Wales. After it French men and women seized and kissed the hand of Louis with the cry, *'Vive l'Empereur!'* But Louis sadly replied, 'No, the emperor is dead. *Vive la France!'* And until the death of Louis the British people continued to call him the Prince Imperial. Queen Victoria felt deeply with Eugénie and Louis over their bereavement. Nevertheless she wrote to a friend: 'The Queen does not think the Bonapartist cause will lose by the poor Emperor's death. On the contrary, she thinks the reverse. For the peace of Europe she also thinks it would be best if the Prince Imperial was ultimately to succeed.'

Louis was aged nearly nineteen when in January 1875 he sat for his final examinations at The Shop. He did well and was placed seventh in order of merit among those cadets passing out. 'Very good—you could not have done better,' commander-in-chief the Duke of Cambridge told Louis at the passing-out ceremony, proudly watched by the Empress Eugénie. And that night at the end-of-term ball the cadets spontaneously seized 'the P.I.', as they called him, and carried him around the hall amid cheers.

Now the Prince Imperial was at Aldershot, where he was introduced to us at the start of this chapter, as a subaltern in E Battery under the watchful eye of tall, moustached Lieutenant Arthur Bigge. He was lonely at first because of the permanent return to France of Louis Conneau, who had been his constant companion since infancy. But he formed a friendship with Bigge which was swiftly to become close and deep. The prince also became friends of two young officers who were special comrades of Bigge, on and off duty. These were Captain F. C. Slade and Lieutenant J. H. Wodehouse. The four were soon to become a little gang, sharing one another's social life and even clothing. They worked together, went on holidays together and got drunk together. The golden years of Louis began at Aldershot and lasted almost to the end of his brief and ill-fated life. He seemed to enjoy work and play with equal zest.

Second Lieutenant the Prince Imperial daily spent hours with the gun teams in his initial training. He became an expert with the nine-pounder muzzle-loaders. Quickly he mastered the techniques of all positions in gunnery. Battery commander Major Ward Ashton, observing the second lieutenant, was deeply impressed by his progress. Other officers at Waterloo Barracks, where the field batteries were stationed, had made cynical remarks about 'this Frenchman' being imposed upon them, but now he was accepted by all with respect.

The Prince Imperial's first public appearance, as it were, at Aldershot was in a great parade on June 28, 1875. On this occasion there were 20,000 men on parade before the Prince and Princess of Wales, the Sultan of Zanzibar and the Empress Eugénie, proud and smiling. In the following month came the Aldershot manœuvres in pouring rain that never ceased. While night after night water dripped through a tent on to the body of Louis in his camp bed, his mother was 'taking the waters' at a spa. He wrote to her: 'I find myself very well in this life—a complete novelty to me. For the five days we have been in the tents it has done nothing but rain. Two days ago the water invaded our canvas home during the horror of darkest night. I found myself with my comrades in the middle of a little lake, in which our belongings were floating about in a wretched state.' If these military exercises offered no excitement or dangers then Louis made these for himself. In the manœuvres of 1876, for example, he insisted upon jumping over camp fires instead of going around them. In one such leap he fell into the fire and badly burned an arm, necessitating his removal to hospital.

In the brigade mess this Frenchman's *joie de vivre* became unbounded on those nights when the rules of etiquette were relaxed. He looked forward particularly to its guest nights when it was permissible to go mad. He would sedately retire from the table and return a few minutes later in the costume of a gymnast. He would leap from a table to a chandelier and then manipulate his body until he was hanging upside down from his feet. He would stand on his head, ask for a glass of wine and drink it. He would try to run up the walls of the mess with the object of touching the ceiling with his fingers. There was the evening when a waiter entered the room with a large tray filled with glasses of water. The prince took the tray from the waiter, placed it on his head. While performing this balancing feat he was close to the chair of Colonel Whinyates at the head of the table. A young officer jumped on to his own chair and gave a high kick, striking the tray. The glasses and water fell on the head of Whinyates and the guest sitting next to him. The colonel expressed the opinion on the following day that the Prince Imperial could at times 'go too far' with his mess gymnastics and that in future he must be watched and curbed. In other after-dinner expressions of *joie de vivre* the prince and his comrades would suddenly start rending furniture to pieces, throwing ruined armchairs and sofas out of the window. This was foolish, because they invariably had to pay compensation for their destructive fun. But this was in an era when there were few officers without substantial private means.

The training of gunnery officers was now considerably more efficient than it had been before the Crimean War. But once the officers knew their jobs and made appearances in some training periods and manœuvres, they were free to spend long periods away from their units. Since in the army of the nineteenth century all officers were believed to be gentlemen, it was appreciated that gentlemen had duties in society, on the field of sport and in foreign travel. Thus their long absences from their units and barracks were understood and willingly permitted by the high command of the army and the War Office. Some officers limited their service with the regiment to less than 120 days a year. In their absence the regiment functioned in the presence of one orderly officer. Senior N.C.O.s did all the serious work of drilling and training the troops. The clubs and drawing-rooms of London, the hunting field and the fashionable resorts of the Continent, always abounded with army officers absorbed in the business of being gentlemen. Because of his own exalted social position the Prince Imperial could disappear from his battery for as long as he chose, and apparently his comrades Bigge, Slade and Wodehouse were free to join him at any time.

Immediately after the 1875 manœuvres this Aldershot quartet were off to Switzerland to stay with the Empress Eugénie in Arenenberg, the villa overlooking Lake Constance, left to her by Napoleon. Once the home of Napoleon's mother, Queen Hortense of Sweden, it was furnished and decorated with stuffy formality. The arrival of Louis and his three friends, together with other young guests, was followed by gaiety and laughter that shocked the villa's ancient servants. These could not hide from the prince's valet their disappoval of the uninhibited antics and conversation of the younger generation. The young ladies and gentlemen of the house party, unrestrained by their royal hostess, played hide-and-seek among the heavy draperies, yellowing statues and massive furniture of the rooms. They giggled at the formal portraits of past members of the Bonaparte family. The meals at Arenenberg became as informal as the picnics. All dignity seemed to have deserted the villa except for the strict protocol still observed in the kitchens and the butler's pantry.

The indefinable charm of the foreign prince which won him the affection of the British army also captivated British society. Certainly Louis was no Apollo. As noticed by Queen Victoria in his boyhood, he was still rather stumpy about the legs. He wore his black hair too short and parted it in the centre. He had a small military moustache. Probably the redeeming feature of the looks of the Prince Imperial were his blue eyes. They were deep and dreamy. They suggested that behind the façade of the efficient part-time soldier and the popular part-time playboy there dwelt a mystic and a poet. But definitely he was no example of a polished Victorian gentleman. He paid no meticulous attention to his dress. His conversation would have an unapologetic loudness. The politician Charles Dilke, meeting Louis at the parties of the Countess Waldegrave, found 'his appearance vulgar and manners common'. Nevertheless he became the darling of British royalty and society. Queen Victoria adored him. To her he was always 'that dear boy'. While even Victoria's own children were a little frightened and diffident in her presence, Louis was as undisturbed as he would have been in the presence of any ordinary woman. The two loved to converse together. The queen's secretary Sir Frederick Ponsonby said that only two people had absolutely no fear of her and these were John Brown, her personal Scottish servant, and the Prince Imperial. 'Does the queen terrify you?' the prince was asked after a meal with her at Osborne. 'Good heavens, no,' was his reply. 'Why should she? We like each other.' The queen's daughter Princess Beatrice also became a close friend of the prince. Their friendship led to recurring rumours of a romance.

Edward Prince of Wales brought Louis into his social coterie known as 'the Marlborough House set'. It was through this that Louis became friendly with the lovely if notorious Lily Langtry and rode with her in Rotten Row. They met often at country house parties where Lily became convulsed by his tricks and practical jokes. He scared highly strung people with stories about the house being haunted. Then later, covered in a sheet, he glided along the corridors on roller skates throwing some who saw him into paroxysms of terror. He thought it funny to enter the bedroom of a woman guest in her absence and place on her bed a top hat and pair of trousers. Next he would go into the room of a male guest and place on his bed a pair of female corsets. The woman, on returning to her room, would fear that a man must be hiding there. And the man, seeing the corsets on the bed, would be equally perplexed. Some victims of this hoax believed they had entered a room other than their own, and would thereupon frantically open other bedroom doors in the corridors. When together at a week-end house party Louis and the Prince of Wales planned a surprise to baffle and annoy a fellow guest. They obtained a small donkey, hoisted it to the window of the bedroom of this man and swung it in. Then they dressed up the donkey and persuaded it to settle on the bed to await his entrance.

Many British people adored Louis, preferring him to their own portly, bearded Prince of Wales. He did not look glamorous, yet they invariably associated him with the glamour and excitement of fashionable international society. When Eugénie took a house at Cowes for the yachting season crowds would wait for hours at its gates hoping for a glimpse of the Prince Imperial. As he moved about London he was constantly being recognised and politely applauded. However, the reception was more boisterous when, late one night, Louis and his friend Arthur Bigge were leaving the Café Royal. He was recognised by a group of these French prostitutes who at that time practised their profession in London in great numbers. The girls loudly cheered their compatriot. '*Vite, vite, mon prince!*' they cried, as he and Bigge fled in embarrassment.

The future of the Prince Imperial seemed to be growing in promise because of the increasing number of people in France who hoped to see this heir to Napoleon III on the throne in a democratic monarchy that would replace the unhappy republic. On his eighteenth birthday in 1874, when he officially became of age, thousands of loyal Bonapartists crossed the Channel to hail him at celebrations at Camden Place. And in the years that followed the Bonapartists watched the progress of the prince, always keeping in close touch with him and with Eugénie. Even during the prince's busiest periods of army training at Woolwich and Aldershot he was in perpetual contact with the Bonapartist leaders.

As for the prince's gaiety and love of fun, these supporters loved him for it. These were, they said, traits of a character which was truly French. They dreamed that soon he might make a grand marriage that would produce fine children for the continuity of the dynasty after its restoration. However, despite speculation regarding his friendship with Princess Beatrice, despite vague unproven stories of his liaisons with other women, Louis had developed no marital ambitions, and he was no sexual philanderer. He was a moral, idealistic young man who, while moving among high society folk of questionable morality, was strict in respecting the moral teachings of the Roman Catholic Church of which he was a devoted member.

The hopes for the future of the Prince Imperial were high. But in 1879, when he was but twenty-three, he took a step that doomed them all and, indeed, ended for all time the prospects for a Bonapartist restoration. The prince took this step because of his attachment to the British army and to the military comrades he made in Aldershot. The cause of it all was the Zulu War, which in the New Year of 1879 at Aldershot was being looked upon as one of those little conflicts which so frequently engaged British troops and which they would inevitably win after a few gallant charges. The facts were that in South Africa an impudent King Cetewayo of the Zulus was refusing to give up his army of warriors known as the *Impis*. Thus General Lord Chelmsford was ordered to lead an expedition into Zululand to punish this defiant native and to disperse the warriors. No one doubted but that Chelmsford and his men would succeed without much effort. Indeed, in the early days of 1879 the troops and people of Aldershot were thinking of other things.

The officers of The Camp held New Year parties where the consumption of wines and spirits were said to have been enormous and extravagant. And in 1879 the price of a bottle of Scotch whisky had risen to 2s. 3d. and of Gilbey's gin to 2s. It was impossible to get a fair sherry under 1s. 3d. a bottle, while a shilling was being demanded for a bottle of claret. Many of the camp's N.C.O.s, with their wives or girl friends, spent the New Year's Eve dancing quadrilles at the Lord Campbell in Alexandra Road, Aldershot, under the direction of a courtly M.C. wearing full evening dress with white gloves. Other soldiers helped to fill Aldershot's Victory Theatre for the new pantomime *Tom, Tom, The Piper's Son*. Also there were crowds of soldiers and civilians around the Rat Pit, then the great attraction of the Royal Military Hotel. They laughed as they watched scores of screaming rodents, dumped into the pit from sacks, being savaged by terriers. But the town did not forget her poor. On New Year's Day 100 gallons of soup were dished out among a horde of the hungry and the needy. The popular local clergyman the Rev. T. Hadow attended the soup distribution. Tasting some himself, he was reported to have 'smilingly pronounced it very good'. On the same day sports lovers flocked to the meadow adjoining the Old Ford Inn to watch a live pigeon-shooting competition for the prize of 'a big fat hog'. Also people with no teeth or those interested in technical progress crowded premises leased for a few days by Mr. B. L. Moseley, a dental wizard from London. He extracted teeth while patients were under 'nitrous oxide gas', the modern wonder that banished pain. He replaced these with dentures which had 'a complete absence of all springs and wires' and which were 'astonishing for the retentive power produced'.

Towards the end of January, however, there was tragedy in South Africa. Details of this, reaching England in the following month, set Aldershot on fire, causing all normal amusements and trivialities of the day to recede into the background. A vital

G

portion of Lord Chelmsford's expedition, marching jauntily into Zululand to 'punish' the defiant chief Cetewayo, had been wiped out. This was a heavily protected supply column, the guards of which were camping in a valley at Isandlwana when 20,000 of Cetewayo's warriors attacked. Some 800 British soldiers and 500 Africans were killed. All were disembowelled by the triumphant enemy. Many officers and men belonging to the 24th Regiment were well known in Aldershot and had left their wives and families there. Immediately in Aldershot a public fund was opened for these dependants. And in The Camp troops were hurriedly mobilised to speed to South Africa on a mission of righteous vengeance. In a sermon at the Iron Church in South Camp, chaplain the Rev. W. H. Bullock told his hundreds of uniformed listeners: 'You are going to avenge your comrades and I am sure the victory will be yours —providing that you go forth in the spirit of God.' Among those preparing to 'go forth' was E Battery of the 24th Brigade.

The humiliation suffered by British arms at Isandlwana rebounded upon Louis Bonaparte, the Prince Imperial. He was advised that he could under no circumstances return to E Battery and that, indeed, it would be advisable for him to keep away from Aldershot while so many of its troops were preparing to depart for war. The prince, despite his feelings for France and for his compatriots there who wanted him back, now felt part of the British army and was outraged by official insistence that he must not return to it. Then, to complete his humiliation, Bigge, Slade and later Wodehouse came to bid him good-bye before leaving for South Africa. Except perhaps for these old battery cronies, all the other friends of Louis— French and British—regarded his idea of joining in the fighting in Africa as preposterous. Frenchmen warned him that if he threw away his life fighting in a foreign army all hopes of a Bonapartist restoration would be shattered. Englishmen explained to Louis that to have him, an important and contro-versial French royal personage, in the army in Zululand would

not merely be a responsibility but also an embarrassment to the
British government. Nevertheless Louis wrote a letter to
commander-in-chief the Duke of Cambridge begging him for
sanction to serve in Africa. Back came a polite refusal. The
empress, who watched her son read the letter, was later to say
that this was the only time since his childhood that she had seen
him cry.

The rejected lieutenant of E Battery replied to the
commander-in-chief with a letter of Gallic subtlety. He wrote:
'I looked upon this war as an opportunity for showing my
gratitude towards the Queen and the nation in a way that
would have been after my own heart. I hoped that it would be
in the ranks of our allies that I would first take up arms. Losing
this hope, I lose one of the consolations of my exile. I remain
none the less deeply devoted to the Queen.' The Duke of
Cambridge, who was a sentimental man, was moved by this
letter from a young officer he greatly admired. The letter was
followed by a personal appearance of the empress in Whitehall
where she confronted the Duke of Cambridge and pleaded
that her son be allowed to follow his army comrades to Africa.

The duke suggested a compromise. This was that the Prince
Imperial should be allowed to go to Africa as an ordinary
traveller, and perhaps he might then be attached to the staff of
Lord Chelmsford in some capacity where there was no chance
of his life being endangered. But first, added the duke, there
must be consultations with both Queen Victoria and with the
Earl of Beaconsfield, the prime minister. The duke, in asking
the queen for her views, showed her the Prince Imperial's letter.
'I am greatly touched,' she said, in agreeing to the suggestion
of the commander-in-chief. 'But Louis *must* be careful not to
expose himself unnecessarily, for we know he is very venture-
some.' Beaconsfield also concurred. When others warned the
prime minister that the scheme was unwise and dangerous he
shrugged his shoulders and sighed, 'I can't stop two obstinate
old women.'

Hundreds stood cheering at the dockside on February 27, 1879, when the empress saw her son off to Africa. She was kissing him and begging him to be careful when a telegram from the queen was placed in her hands. It told Eugénie that her 'beloved son' was leaving 'with the good wishes of the entire nation', and ended with the words: 'May God bless and keep him.' On arrival in Africa, Louis was appointed aide-de-camp to army commander Lord Chelmsford. Now, very near to his final destiny was this boyish mystic who had written: 'If one belongs to a race of soldiers, one finds recognition only with sword in hand, and he who wants to learn by travel, must travel far.'

Louis was, as arranged, 'attached' to Lord Chelmsford. But he was too bold and too slippery to vegetate in a place of safety. Constantly he was courting danger with units penetrating into enemy territory. At times he became too reckless, too exuberant and something of a show-off. By such behaviour he even offended the British army's greatest exhibitionist warrior, Redvers Buller, who is fully discussed in Chapter 7. Louis was with a column of Colonel Buller's regiment of horse when sighted on a hill-top were a mass of enemy Zulus. When the horsemen charged up the hill towards the Zulus, Louis spurred on his horse until he was in the lead. The Zulus ran away. Then Louis galloped around at the top of the hill giving loud 'Tally-ho's' as if at an English hunt. Buller was annoyed at such conduct from the prince, regarding it as vulgar and ungentlemanly. He complained to Lord Chelmsford about it. Nevertheless Louis was generally very popular among younger officers and men. His standing in the eyes of Lord Chelmsford was also high. 'I think of him,' said the expedition's commander-in-chief, 'as one of the most reliable officers on my staff. He has my full confidence. I congratulate myself on having him with me—and I know he wants to be regarded exactly as my other officers.'

Next Louis, still evading safety, became assistant staff officer to Colonel Harrison, the Quartermaster General. A big advance was being planned against the Zulus. Reconnoitring parties were being sent ahead to survey the ground and to find suitable camping spots for the troops who would be coming. One night Harrison ordered Louis to join such a party on the following morning. This was to be in charge of a Colonel Bettington, a reliable and trusted officer. It was to consist of six white and six Basuto scouts. There were last-minute changes which were to prove fatal. Bettington was reassigned to other important duties. His place was taken by Lieutenant J. Brenton Carey, an officer of pleasant but excitable disposition. Due to a misunderstanding only six white scouts and no Basutos appeared for the mission. As this depleted party was leaving, Colonel Harrison, in wishing them well, said to Carey: 'Remember—you will look after the prince.' Another officer said to Louis: 'Take care of yourself, and don't get shot.' 'Oh no,' was the reply. 'Lieutenant Carey will take good care that nothing happens to me.'

The party moved into the wilds, Louis astride his big grey horse named Fate, which he had brought from England. Later, in the high grass, Louis and Carey were discussing the campaigns of the great Napoleon. Suddenly the head of a Zulu appeared above the grass a short distance away. Carey leaped on to his horse and rode off. Now there were at least thirty Zulus advancing and firing rifles. The prince tried to mount Fate, but the horse, always nervous, bucked and reared, causing its master's sword to fall on the ground from its scabbard. The six scouts were on their horses and one cried to Louis: 'Get mounted, sir. For God's sake get mounted.' But Fate galloped away towards the riding scouts. Louis ran after his horse and, in making another attempt to vault into the saddle, fell on his face, crushing his right arm.

Louis rose. With the left arm he drew his revolver. He walked towards the enemy and fired the only three bullets that were in the revolver. The body was found next day. It had seventeen spear wounds, all in the front, and it was stripped of clothing. But still hanging on a chain around the prince's neck were the religious medallions he always wore. Some of those Zulus who had killed Louis were later captured. They explained that they had not disembowelled him because of this 'magic chain' which they dared not remove. Even these warriors had been touched by the courage shown by the prince in a desperate resistance against impossible odds.

The body was from the Cape placed aboard H.M.S. *Orontes* bound for England. Travelling with it was broken-hearted Captain Arthur Bigge, whose meeting with Louis at Aldershot had led to such deep brotherly friendship. However, the news of the tragedy did not reach England until June 19—eighteen days after its occurrence. Since there was no cable laid to the Cape, the report of the death of the Prince Imperial had to be carried by sea to Madeira. From there it was cabled to the Duke of Cambridge in London. The duke telegraphed Balmoral, where Queen Victoria was in residence. The queen's diary tells how in the morning her faithful servant John Brown knocked on the bedroom door and told her, 'The young prince is killed.' The diary continued: 'I put my hands to my head and cried out, "No! no! It cannot, cannot be true. It can't be." Dear Beatrice, crying very much, as I did, gave me the telegram.' Lord Sydney, a member of the queen's entourage, was sent by overnight train to London and then to Chislehurst to break the news to Eugénie. Gently, Lord Sydney told the empress her son had died gallantly in action. She screamed and fell in a long, merciful faint.

Aldershot, and Woolwich too, went into mourning. Shop-keepers nailed strips of black painted wood over their show windows as symbols of commercial grief. In many Aldershot shops the windows were emptied of merchandise to be replaced by a portrait of the Prince Imperial, the frame swathed in crape. There were memorial services to the Prince

Imperial in churches of all denominations. Crowds of Roman Catholics filled their Aldershot Church of St. Michael and St. Sebastian for several days and nights praying for the Prince Imperial's soul. And in the mess of the Royal Artillery officers recalled how Louis ran up its walls and performed acrobatics from the beams. They buried Louis with black Victorian splendour in St. Mary's Roman Catholic Church, Chislehurst, where his father Napoleon III also lay. Queen Victoria was there, weeping with Princess Beatrice, and with a tribute of white globular peonies. The pall-bearers were the Prince of Wales, Prince Arthur Duke of Connaught, Crown Prince Gustave of Sweden, a Monsieur Rouher and the Duc de Bassano. The procession that preceded the internment was led by the Royal Artillery Band with muffled drums. The coffin of the Prince Imperial rested on a gun-carriage. His horse Stag was almost smothered in black and purple funeral draperies.

When the Empress Eugénie and others went through the personal possessions of the dead Louis they discovered that behind all his laughter and gay exhibitionism there was a character of deep spirituality and earnestness. For example, as Eugénie was glancing over one of his devotional books, out fluttered a piece of paper bearing a prayer in the handwriting of her son. 'O my God,' it said, 'show me always where my duty lies; give me the strength always to do it. Instil deeper into my heart, O God, the conviction that those whom I love and who have died are witnesses of my actions. My life will be worthy for them to see, and my inmost thoughts will never cause them to blush.' Then in his last will and testament, written the night before he sailed for Africa, Louis revealed a premonition of his early death. 'I shall die', said this will, 'with a sentiment of profound gratitude to Her Majesty the Queen of England and the Royal Family and for the country where I have received for eight years such cordial hospitality.' The prince's last bequest read: 'To my mother, I leave the last uniform I shall have worn.'

In consequence of the death of his great friend, Arthur Bigge entered upon a career of which he had never dreamed. When he visited Queen Victoria to tell her about the last days of Louis in Africa she was so impressed by him that she invited him into her personal service. So this son of a country vicar abandoned the army. He became trusted private secretary not only to Victoria but also to Kings Edward VII and George V. On being honoured with a peerage, he chose the title of Lord Stamfordham, after the village of which his father had been vicar. As for the cowardly Lieutenant Carey, who ran away when the Prince Imperial faced death, he was court martialled, found guilty and sentenced to be cashiered. However, because of cleverly organised pressure by his friends, the case of Carey was reopened and the strong verdict against him was rescinded. Instead he was given a severe reprimand and was allowed to return to his old rank and regiment where he suffered much cold shouldering and many snubs until his death in India.

The body of the Prince Imperial did not remain in Chislehurst. In 1880 the Empress Eugénie moved to Farnborough, close to Aldershot. Near her new home she endowed an abbey. There in the crypt were placed the bodies of Louis and his father, to be joined by that of Eugénie when she died so many years later in 1920 at the age of ninety-four. Many soldiers come from Aldershot, where stories of the fun-loving Prince Imperial still linger, to see his grave and that of his parents. And some of these visitors insist that the cold silence of the crypt has been broken mysteriously by a gay ripple of laughter as if the spirit of the Prince Imperial were striving to make his presence known.

5

The Statue

WHEN The Camp at Aldershot was founded in 1854 the Duke of Wellington had been dead two years. Nevertheless his name has a big place in local history because of what Victorians called 'the extraordinary affair of the Wellington statue'. The statue of this 'affair' stands on a lonely mound called Round Hill, near the Farnborough Road at Aldershot. It is a huge bronze statue, the uniformed duke in his short cloak and feathered helmet gazing reproachfully towards the nearby house of the G.O.C., from the back of his favourite horse Copenhagen. Around are sixteen old cannon barrels, dirty and rusty, sunk in the ground. The slopes of the mound are rich with silver birch and fir trees, rhododendrons and other bushes. The grass at the plinth, however, is rough and rarely cut. The statue's vital statistics are impressive. The girth of the horse's body is 22 ft. 8 in., the girth of its legs 5 ft. 4 in. Copenhagen measures 26 ft. from nose to tail. Its head is 6 ft. and each ear is 2 ft. 2 in. The horse weighs 33 tons and its rider 7 tons. Today only a few people look at them, despite their size. However, couples who favour the surrounding bushes for love-making will sometimes climb to the plinth to scribble such information as 'Mat loves Agnes' or to write their names inside a badly drawn heart. Then birds in great numbers use the statue for repose and for intestinal relief.

The Wellington statue is an exile. That is probably why its appearance seems alien to the Aldershot scene. Unlike other local memorials, the churches and the barracks, it was not made for Aldershot but simply dumped there in 1885 in a ceremony that seemed insincere and false. Then the size of the statue makes it look as incongruous as Gulliver looked in Lilliput. Duke and horse were, indeed, fashioned in proportion to the greatness of London where, until 1885 they stood gloriously on a triumphal arch at Constitution Hill, not far from Buckingham Palace itself.

The colossal statue was created during the first half of the 1840s when the Duke of Wellington, then in his seventies, was repeatedly and bitterly complaining about the number of artists and sculptors demanding that he sit for them. Since the victory of Waterloo in 1815 portraits, sculptures and also vast panoramic action paintings of Wellington had been pouring from the studios of England and the Continent to supply a demand that seemed insatiable. Every week Wellington received a mass of letters from artists and sculptors begging him to sit for them. Stung to anger by such 'persecution', the duke wrote: 'I can positively sit no longer. I do think that having to pass every leisure hour that one has in daylight in sitting for one's picture is too bad. No man ever submitted to such a bore; and positively I will not sit any longer.'

The artist Benjamin Haydon, beset by debt, decided that he might return to solvency by making a picture of Wellington and selling it at a handsome fee. Wellington at first refused to grant him any sittings. Thereupon in desperation Haydon wrote to the duke asking if he could borrow one of his uniforms with accoutrements. Haydon apparently reasoned that the clothing would be authentic, even if the head, hands and contours of the ducal body had to be copied from existing portraits. But Wellington's cold reply was that 'The Duke knows nothing about the picture Mr. Haydon proposes to paint. At all events, he must decline to lend anybody his clothes, arms and equipment.' Nevertheless the duke eventually relented and Haydon was allowed to stay with him at Walmer Castle and paint him from life.

There was a City of London councillor and business man

named Thomas Bridge Simpson whose admiration for Wellington amounted almost to a passion. Simpson agitated for the erection of an equestrian Wellington statue that would 'dominate the City'. He suggested that funds for the statue be raised without delay and that the sculptor Matthew Coles Wyatt should be commissioned to execute it. For Wyatt had been the creator of the fine statue of the mounted George III which still stands in Cockspur Street near Trafalgar Square. A committee was duly assembled in the City and money for the statue soon guaranteed. But there was a battle among those members who, like Simpson, wanted Wyatt to execute the statue and those who favoured the sculptor Sir Francis Chantrey for the task. Chantrey had fashioned huge busts of Howe, Vincent, Duncan and Nelson for the Greenwich Hospital. And among his statues was one of King George III in the Guildhall and of the king's arch-enemy George Washington in Boston, Massachusetts.

The champions of Chantrey were victorious. But Chantrey died in 1842 before his equestrian Wellington for the Royal Exchange was finished. Another sculptor named Weekes had to complete the job. Meanwhile Mr. Simpson felt that a Wellington statue for the City only was insufficient. He was moved by a vision of a colossal bronze Wellington and horse which, raised in the West End of London, would forever awe and inspire all who passed by. Mr. Simpson's dedicated enthusiasm led to the formation of a new body known as the Wellington Military Memorial Committee with the politically influential seventh Duke of Rutland as chairman. A sum of £14,000 was raised and, to Mr. Simpson's satisfaction, Matthew Coles Wyatt was given the commission for the statue.

In 1828 a Triumphal Arch had been built by public subscription at the Hyde Park Corner entrance to the Buckingham Palace domain. Some members of the committee and others suggested that the ideal place for the planned giant statue would be on top of the arch, although it was admitted that people

would have to crane their necks to look up at it. Sculptor
Wyatt became angry. He argued that Wellington, on such an
elevation, would look absurd. There were acrimonious arguments about the matter in the Houses of Commons and Lords.
It was asked that if, as planned, Nelson could be on a column
170 ft. high,* why should not Wellington be on the arch?
Finally, the opposition of the sculptor and others was dismissed. It was resolved that the equestrian statue should go on
the arch, it being pointed out that the living Duke of Wellington would be able to see it from the windows of his mansion
Apsley House which was nearby.

Wyatt first made an experimental model in wood. This was
hoisted to the top of the Triumphal Arch to give the Wellington Military Memorial Committee an impression of what the
real thing would be like. The general feeling was that the effect
of the giant horse and its giant rider at such an elevation was
impressive. Next Wyatt gingerly made moves to approach
Wellington to pose for him. But, alas, the duke had become
still more irascible as the demands from the world's artists and
sculptors to give him sittings increased rather than decreased.
On being told that he should, for diplomacy's sake, pose before
a foreign royalty for a marble bust, the duke cried: 'Damn
him!' and then acquiesced. Of course, the duke realised that to
refuse all co-operation with Wyatt for the equestrian statue
would offend the many distinguished people who were
contributing towards its creation. Therefore he grudgingly
allowed Wyatt to come to Walmer Castle to make a few
drawings from life—but that was all.

Wyatt and his son James started work on the statue in the
early summer of 1840 on a huge turntable fixed to forty
rollers, enabling it to revolve at their convenience. Father and
son laboured from a movable platform that could be adjusted
to various heights. The model, on which three tons of plaster
of paris was used, took three years to complete. It was then

* Erection of the column started in 1840.

cast in bronze in a foundry made for this one statue. Because of
its size, the statue had to be cast in pieces, later to be brought
together with screws and with metal fusion. Meanwhile in the
Royal Naval Dockyard at Woolwich an army of workers
were making a 20-ton dray, described as 'a carriage' to give it
glamour. This—the biggest dray ever constructed—was to
convey the statue in triumphal procession from the Wyatt
studio and foundry in Harrow Road to the Triumphal Arch.

London welcomed the bronze Wellington and horse on the
sunny morning and afternoon of September 29, 1846, as they
would have welcomed some fabulous and popular foreign
royalty on a state visit, or their own ruler on her way to some
great national ceremony or celebration. Stands for spectators
were built from Harrow Road, into the Edgware Road and
down Park Lane to Hyde Park Corner and the Triumphal
Arch. Shopkeepers and even private residents rented seats in
their upper windows to people who had come from all over
the country for the occasion. Troops and police lined the
streets to control the crowds. At 9 a.m. the commander-in-
chief, George Duke of Cambridge, arrived at the Wyatt
studio amid loud applause. George, accompanied by the Grand
Duke of Mecklenburg-Strelitz, had come to inspect the statue
before its début. As the clocks of the metropolis struck ten, the
dray with statue moved out of the Wyatt studio to the hurrahs
of the waiting thousands. It was being dragged by 100 sweating
soldiers of the Scots Fusilier Guards. Then twenty-nine horses
were harnessed to the dray. These were beautifully groomed
and had been borrowed from a firm of brewers. On their
heads were leaves of laurel, symbols of victory.

Two hours passed before the procession began to move. In
the lead were mounted men of the 2nd Life Guards. They were
followed by the bands of three guards regiments, behind whom
marched 500 more guardsmen. Then came the statue on the
dray, with an escort of mounted troopers on each side. Behind
were the 100 of the Scots Fusilier Guards and several score

workmen who had helped to make the dray in the Woolwich dockyard. Also there were rows of marching gentlemen in frock coats and top hats, representing the Wellington Military Memorial Committee and other important bodies. More mounted men of the 2nd Life Guards trotted at the rear of the procession which moved down the Edgware Road to the tune of 'Here the Conquering Hero Comes'. On stands at Hyde Park Corner were officials from Buckingham Palace, bishops, noblemen, members of parliament and other dignitaries to greet the statue and to watch it raised by various mechanical devices up a 115 ft. scaffolding to the top of the arch. The statue reached the scaffolding at 2 p.m., but roping it up for haulage to its lofty destination proved so difficult that operations were postponed for the day. Indeed, forty-eight hours had passed before the statue was safely on its perch.

Yet, after all the ceremony and enthusiasm and expense, the position of the statue almost immediately aroused criticism. Thousands of people came to see the massive sculpture and it became recognised as one of the sights of London. Nevertheless critics said it made the Triumphal Arch look top-heavy, that it was cumbersome and incongruous in such a place. *Punch* and other satirical magazines made fun of it. Some art critics damned it with such words as 'gross' and 'monstrous'. From the denunciations arose agitation for the statue's dismantlement and banishment. The matter became a bitter, controversial topic in the House of Commons and House of Lords. There was correspondence in the newspapers attacking and defending the statue. The prime minister, Lord John Russell, joined the statue's opposition. Finally, it became known that Queen Victoria herself was objecting to the monster horse and its rider towering almost in the shadow of her palace. Then came news that the deaf old Duke of Wellington in nearby Apsley House was angry and pained over the moves to banish his image and horse from the Triumphal Arch, and that he regarded these as disrespectful and insulting to his person. So great was the prestige of Wellington that the nation behaved like a group of naughty Victorian schoolboys rebuked by their stern headmaster. Opposition to the statue evaporated. The critics slunk away in silence and in shame. The statue remained sacrosanct on the top of the arch.

In 1852 Wellington died. And as the years passed, it seemed that the bronze Wellington and Copenhagen at Hyde Park Corner would remain permanently as one of London's landmarks. However, in 1882 they became under assault again. Much of west and central London was being replanned. Road space at Hyde Park Corner was to be increased by utilising a portion of the Green Park and Constitution Hill. For this to be achieved the Triumphal Arch with its statue would have to be removed. Wellington had been dead thirty years, but controversy flared up again as champions of his memory demanded that the statue remain where it was. Mr. Gladstone, who was now prime minister, supported the decision of his Commissoner of Works, Mr. Shaw Lefevre, that it should be moved.

Some of those who reverenced the statue said that the magazine *Punch* should be publicly burned for continually treating it with levity. Particular objection was taken to *Punch*'s issue of April 8, 1882, which added fires to the controversy with three items—a full-page cartoon, a satirical playlet, and verses. The cartoon depicted Shaw Lefevre with a 'Notice of Removal' in his hand, climbing up the statue on a ladder, with Wellington glaring down from the horse. The duke is saying: 'Hey? What? Come down? Why certainly—and don't put me up again.' The playlet was called *Don Shaw-Vanni and the Statue or The Virtuous Lothario and His Happy Thought*. Its three characters were Don Shaw-Vanni (of the Office of Works), Il Commendatore (Il Duco de Ferro) and Leporello (of Fleet Street). The verses suggested to Shaw Lefevre that he:

'. . . take the Duke as he is now,
Fix him hat, horse and all on white Dover's cliff brow.
He'll protect us, for should that new tunnel be made,
Mark my words, 'twill be giving invaders free trade.
Let him watch o'er the Channel as once he would do;
He has kept the Cinque Ports* as he kept Waterloo;
For if Wellington can't—then who can stop the French?'

Strange that such doggerel should have incited champions of
the statue to a white heat of indignation.

The statue then suffered a period of squalid indignity. It was
early in 1883 lowered from the top of the Triumphal Arch
with mechanical jacks to the laughter of crowds of onlookers.
Next it was moved by steam engines to the fringe of Green
Park where it was propped up by unsightly beams. The anger
of the admirers of the statue over such indignities grew more
tense when it became known that the Triumphal Arch was to
be saved and given a new position on the top of Constitution
Hill. And on the arch top, where the statue had stood, a
quadriga (four-horse chariot) was to be placed. However, as a
concession Gladstone, supported by the Prince of Wales, agreed
that a new small statue of the duke should be placed near the
old site of the arch. But protesters demanded: What is to be
done with the old statue?'

There was indeed a plot to exile this discarded Wellington
memorial from London. Taunted by Punch and other papers
for allowing the statue to be dumped indefinitely in Green
Park, Commissioner of Works Shaw Lefevre and fellow con-
spirators decided that the controversy would soon be forgotten
if the giant equestrian statue were presented to some provincial
town where it might be greeted with gratitude and approval.
Anyhow, it was felt, what provincial mayor would dare look
such a big gift horse in the mouth? But Mr. Shaw Lefevre was
much relieved when the Prince of Wales formed a committee

* Wellington had held office as Lord Warden of the Cinque Ports.

Left Albert, the Prince Consort. Painting by F. Z. Winterhalter, 1859
Right 'The Great German Sausage'. Field Marshal His Royal
Highness George 2nd Duke of Cambridge, commander-in-chief of
army, 1856–95. Painting by Sir Arthur Cope, R.A., in United
Service Club

Above Mrs. Louisa Daniell
Below Mrs. Daniell's Soldiers' Home and Institute

Above Photograph from the satin edition of the *Aldershot News*
printed to honour the kaiser on his visit to Aldershot and The Camp
in August 1894. The kaiser is wearing the British uniform which
disappointed the crowds
Below Queen Victoria reviewing the troops at Aldershot

THE RAT-PIT · 1890 · ROYAL MILITARY HOTEL ·

Above The four-year-old Prince Imperial, in the uniform of the Guards, entertains young friends while attending army manœuvres in the Bois de Boulogne on 30 November 1860. The prince's father, Napoleon III, is the third of the four standing figures on the left. Painting by Adolphe Yvon

Below The Rat Pit, Royal Military Hotel, Aldershot. A contemporary drawing

COMING DOWN!!!

STATUE (log.). "HEY? WHAT? COME DOWN? WHY, CERTNLY! AND—*DON'T PUT ME UP AGAIN!*"

"The plan will involve the removal of the Wellington Arch from its present position."—*See Mr. Shaw Lefevre's Speech on the Hyde Park Improvements.*

Right Punch cartoon on the planned removal of the Wellington statue from London to Aldershot *Below* Single-handed, Lieutenant Evelyn Wood routs a body of natives at Sindwaha, India, in 1858. Impression by a contemporary artist

Left General (later Field Marshal) Sir Evelyn Wood, v.c., g.c.b.
Right General Sir Redvers Buller, v.c., g.c.b.

Funeral of the Late Colonel Cody at Aldershot. One of the Van Loads of Wreaths.

Above Aldershot G.O.C. General Sir John French in the army's
first motor car at The Camp in 1902
Below Wreaths and military mourners at Cody's funeral

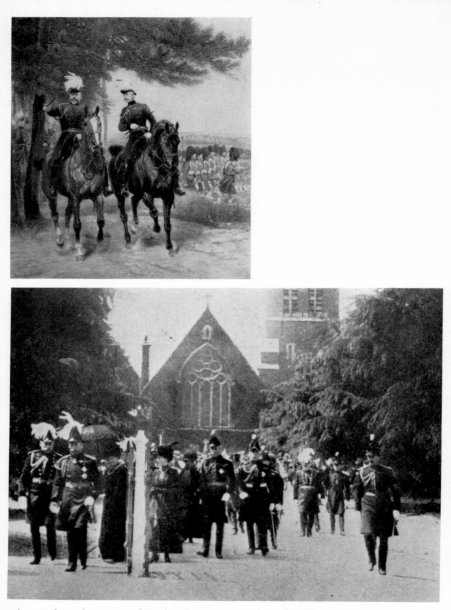

Above Edward Prince of Wales (later King Edward VII) with Prince
Arthur Duke of Connaught (then G.O.C. Aldershot) at Aldershot,
1895. Painting by Edouard Detaille
Below Lieutenant-General Sir Douglas Haig leaving Aldershot's
Garrison Church of All Saints with his friend King George V in May
1914. *L to r*: Haig, the King, Queen Mary, and Princess Mary (later
the Princess Royal)

to advise him on the disposal of the statue. The committee sparkled with aristocracy and influence. It included such figures as Wellington's son, the second duke, Lord Hardinge, Lord de l'Isle and a member of the House of Rothschild. The Prince of Wales who, as Edward VII, was hailed as 'Edward the peacemaker', thought of a perfect destination for the unwanted statue. This was Aldershot, home of the British army and thus an ideal foster home for the bronze image of the army's greatest commander. All the other members of the committee heartily endorsed the prince's suggestion.

The valiant rear guard action of the statue's friends was worthy of that of Wellington's finest troops in his military campaigns. Lord Stratheden and Campbell dared the displeasure of the Prince of Wales and also of the queen by rising in the House of Lords to present a 'humble address' that the statue not be moved from London to Aldershot. 'The government,' said Stratheden, 'has amazed the public and this house by saying they would move the statue without the sanction of parliament. Removal of the statue would, indeed, be an unauthorised encroachment on the rights and enjoyments of the inhabitants of London. By a large and influential section of the world the proposed removal of the statue is regarded as a gross indignity to the memory of a great man and a wanton outrage on a sacred testamentary injunction.' But Lord Sudeley rose in defence of the removal. He stressed that the new Duke of Wellington favoured it and that Aldershot, as the army's greatest centre, would feel proud and honoured to have it. Lord Stratheden's motion was defeated by six votes.

Those who wanted London rid of the statue now expounded a new argument. They suggested that the statue was worthless because Wellington never seriously sat for it. Wyatt may or may not have drawn Wellington from life, they said, but he did most of his work copying from a bust of the duke made by sculptor Joseph Nollekens in 1815. From Belvoir Castle the Duke of Rutland wrote to *The Times* that since it was now

H

proved that Wellington *did* sit for the statue 'reasons for removing it have gone'. When the opposing sides continued to argue whether Wellington did or did not sit for Wyatt, the Prince of Wales quietly pursued negotiations for the statue's exile to Aldershot.

Work on moving the statue and preparing for its transport to Aldershot began at the strange hour of I a.m. on August 7, 1884. For demonstrations against this removal had been feared. Nevertheless even at I a.m. there was a big crowd waiting outside the hoarding which for some time had been erected around the unfortunate sculpture. Guarded by police, work-men began detaching parts of the horse and the body of its rider because it would be easier to take them to Aldershot in pieces. By 10 a.m. the body of Wellington and the head and tail of the horse had been disconnected from the whole. The body was put on a massively built truck with large gun-carriage wheels drawn by four powerful horses. The horse's head, tail and body were moved in a railway trolley pulled by three horses. The journey to Aldershot started. It took nearly four days. The fragmented horse and rider entered Aldershot without ceremony, without bands to play 'Here the Con-quering Hero Comes'. They were dumped in an army storage yard at South Camp.

Next day an Aldershot firm of builders and contractors started, with army assistance, to move the pieces to Round Hill to be reassembled and erected on a new red corshill stone plinth, bearing the name 'Wellington' on either side. Perhaps it was an uneasy sense of guilt over the statue's banishment from her capital that caused Queen Victoria expressly to command that its formal handing over to the Aldershot Divi-sion of her army should be made with a full ceremony and military parade. The Prince of Wales with his son Prince George (later King George V) and Lord Arthur Somerset arrived at the Royal Pavilion, Aldershot, for a preliminary inspection of the statue on its new plinth a day before the presentation ceremony which was on August 19. Then, shortly before the ceremony they were joined by Prince Arthur Duke of Connaught and the Grand Duke of Hesse.

The military welcome to the statue was splendid, troops who included two squadrons of cavalry and a battery of Horse Artillery forming around the plinth. There the Prince of Wales formally 'handed over' the statue to the care of Lieutenant-General D. Anderson, then commanding the Aldershot Division.

'The queen has entirely approved and sanctioned my doing this,' said the prince. 'I feel sure it will be safe in your hands and in those of your successors. I hope the fact of the statue being placed here will cause all soldiers of the garrison to try to emulate the many good qualities of the great duke. I am very glad that my brother-in-law the Grand Duke of Hesse is here today, as representing the German army, and also my brother Arthur, as he was the Duke of Wellington's godson.'

General Anderson effusively expressed to the Prince of Wales the gratitude of all of Aldershot's soldiers for the statue, patronisingly adding: 'We believe that as education extends the duke's noble character will be more and more understood by them.' As massed bands played the National Anthem there began a royal salute of twenty-one guns.

The bronze duke on his bronze horse was thus rehabilitated in Aldershot with every honour. But today they seem lonely, forsaken, out of place. Soldiers of The Camp rarely glance at the statue to be inspired 'to emulate the many good qualities of the great duke'. The soldiers are far better educated than in 1884, yet it remains doubtful whether 'the duke's noble character' is 'more and more understood by them'. At the statue it seems that only the birds remember.

6

The Good Seat

MANY field marshals have, in the glory of their rank and fame, honoured Aldershot with their presence. Yet only one lies buried in the place that advertises itself as 'Home of the British Army'. In its military cemetery, surrounded by a multitude of lesser rank, lie the remains of Field Marshal Sir Evelyn Wood, v.c., under a cross of plain stone. And after the funeral in 1919, attended by three other field marshals, an officer was heard to remark that he couldn't really believe that a man of such fantastic activity could be permanently stilled. Aldershot was also a suitable place for Wood's burial because in his long army career he had spent so many years there. He was at The Camp first in 1866, soon as Brigade Major and then as Deputy Assistant Adjutant General. He returned there in 1876 for two years as Assistant Quartermaster General. From time to time he was there again on a variety of duties. Finally, he was back as General Officer Commanding, from January 1889 to October 1893. Wood's life survived the reign of Edward VII and straggled on into that of George V. Yet he was in spirit and philosophy every inch a Victorian in which age he flourished. He has been described as the last of the great outstanding Victorian military commanders. Modern judgement might delete the word 'outstanding'. But he can accurately be remembered as exponent of the now defunct and disproved theory that military prowess and enterprise were best developed by officers maintaining 'a good seat' on their steeds in the hunting field.

Wood's career as warrior began when he joined the Royal

Navy as a midshipman in 1852 at the age of fourteen. To switch from one service to the other was not an uncommon practice in the nineteenth century. Another famous soldier, General Sir Henry Hildyard, G.O.C. at Aldershot in 1901 and 1902, had served five years in the navy before he transferred to the army. Wood's naval service lasted but three years five months—but in that period he covered himself in glory.

Evelyn was the son of a baronet clergyman, the Rev. Sir John Page Wood, who held plural office as rector of St. Peter's, Cornhill, London, and vicar of Cressing, near Braintree, Essex. Evelyn's education began at Marlborough Grammar School, from which he moved to the public school of that town. Marlborough College was at that time a place of turbulence and cruelty, cursed by a running war between the 500 boys and their masters. When, in November 1851, the headmaster forbade the celebration of Guy Fawkes Day the school rose in revolt, exploded fireworks in the building and fought the masters. Desks of the masters were burned. One boy with a crossbow implanted a pellet in the forehead of the head-master. When the rebellion was eventually suppressed many boys were expelled. The fourteen-year-old Evelyn was spared expulsion but publicly flogged. But he was so popular that the boys turned their backs in protest during the flogging. Then they had a whip-round and presented Evelyn with five pounds. Despite the tumultuous distractions of this unusual public school, the boy was quickly mastering Latin and Greek. However, he wrote to his parents imploring them to help him get away to sea. This resulted, through influential family friends, in a nomination for the Royal Navy.

The little fellow was full of swagger, full of confidence. He went to sea in H.M.S. *Queen,* a 3,000-ton sailing craft. By sheer chance the command of the 116-gun *Queen* passed to Evelyn's maternal uncle Captain Frederick Michell. Michell was a hard disciplinarian. He was particularly strict with Evelyn in order to prove that he treated relatives without favour. When the Crimean War was declared in 1854, *Queen* was ready with other naval vessels in the Black Sea. During the siege of Sebastopol, Wood went ashore with a naval brigade. In the heat of the battle he brought up powder through flames from which others shrank. When a powder magazine caught fire, he climbed on it and stamped out the smouldering flames, burning away the seat of his trousers. His heroism drew the rapt admiration of Captain Peel, commander of the brigade. He became Peel's aide-de-camp and then, after many heroic exploits, was seriously injured in the assault on the Redan. When Evelyn returned to England he was but seventeen, and these were the honours he had earned during his Crimea adventures: mentioned twice in despatches; medal with two clasps; Knight of the Legion of Honour; membership of the 5th Class, Medjidie; Turkish Medal.

No sooner was Evelyn back in England than he was tortured by nostalgia to return to the Crimea and the fighting. Since a naval brigade was no longer there he resigned from the navy and obtained a commission as a cornet in the 13th Light Dragoons, soon bound for the Crimea. But on reaching Turkey Evelyn became ill with typhoid and complications. His mother Lady Wood left for Turkey on receiving letters that he was unlikely to survive. He lay in a military hospital in Scutari in a state of delirium while his mother, gazing at the wasted form, despaired. These were tough days, despite those 'gentle English nurses', stories about whose sweetness and compassion brought tears to the eyes of Queen Victoria. One morning in Scutari Lady Wood quietly opened the door of the invalid's room to see how he was faring. She was just in time to see Evelyn's nurse strike him in the face with her fist for rubbing his sore head during delirium. She was a volunteer from St. Thomas's Hospital in London.

Evelyn survived everything. He certainly missed any more Crimean carnage, but he had not been back in England long before the Indian Mutiny was blazing. One of the big attrac-

tions of the army of the Victorian age was that it never lacked
diversions for those who enjoyed adventure and risks. The
queen's soldiers seemed always to be fighting 'righteous
battles' somewhere against brown or black men. During the
hot summer of 1857 troops poured from Aldershot and other
military centres as news of the Indian rising became progres-
sively graver. Fewer than 40,000 British troops, it was reported,
were desperately and gallantly trying to hold in check a
fanatical population of 100,000,000. Evelyn Wood was transferred to the 17th Lancers and by October 1857 was sailing
towards India, spending hours daily in the study of Hindustani.

In India, Evelyn, now twenty, added lustre to his martial
reputation while commanding native cavalry which pursued a
master rebel named Tantia Topi, damned as 'unscrupulous,
most persistent and elusive'. In confrontations with Tantia's
warriors Evelyn and they carried on a melodramatic dialogue
as they fought. 'Come on you dogs of government!' cried a
Wilayati rebel officer, equipped with sword and shield, as
Evelyn and his men approached. The Wilayati threatened,
'Your body will be food for dogs,' and Wood replied, 'Stop
talking and come on!' Then the Wilayati and his men accepted
the invitation, charging Wood and his faithful orderly Dhokul
Singh. The orderly quickly despatched the defiant Wilayati,
slicing his face in half with his sword. Then the Wilayati's men
surrendered, begging that they might not be slaughtered with
swords. 'Shoot us, sahib, shoot us, please,' they cried. Wood
struck two of the pleaders in their faces with his fist, like the
nurse in the Scutari hospital. At that the men ran away. Next a
friend of Wood named Lieutenant Harding appeared with a
troop of the 8th Hussars. Seeing two rebels who were near
them, armed and defiant, the two officers on their horses
charged them. One rebel hit Wood with his musket, but
Wood deftly pierced him through the body with a sword. And
the men rode down by Harding perished also. For this affair
Wood was mentioned in despatches.

Again Evelyn Wood became engaged in conversational combat while still in pursuit of the rascally Tantia Topi. In the wilds he came face to face with a rebel Sepoy. The Sepoy dropped his gun and unsheathed a sword. 'Sahib,' he cried courteously to Wood, 'I know I must die, but to kill individuals is not an officer's duty. Send a private to fight me. I shall either wound or kill you.' Pulling at the reins of his horse, Wood moved towards him with the reply, 'It is more likely I shall kill *you*.' Stooping groundwards to dodge the sword cuts of his mounted adversary, the Sepoy infuriated Wood by slashing at the hind legs of his horse. 'I took you for a warrior, not a horse slaughterer,' shouted Wood in rebuke. Then some of Wood's men galloped up, surrounded the horse-slashing Sepoy and shot him dead for his unkindness to an animal. To add to the sad perfection of the story, Tantia Topi was eventually trapped and held prisoner. Sentenced to the gallows, he suffered a slow and painful death because his executioners bungled their task, never having hanged a man before.

The deed that won Evelyn Wood the Victoria Cross was when, with a party of only ten of his men, he routed a gang of eighty bandits and rescued their pro-British captive, a village chief named Chemmun Singh. Chief of the gang was Madhoo Singh, unscrupulous and cruel. Guided by an informer, Wood and his tiny force crept towards the rebel lair. There he saw Chemmun being roped to a tree with other captives. 'Who is there?' shouted a rebel sentry. 'We are the government,' replied Wood, who turning to his men gave the order, 'Fire! charge!' Wood and his small force duelled with the eighty bandits who eventually fled. He brought Chemmun Singh back to his village. There Chemmun's comely wife prostrated herself before Wood, kissing his boots and repeating, 'Thank you, sahib,' in a scene anticipating the Hollywood films of the twentieth century.

After still more intrepid exploits in India, Wood returned to England as one of her most publicised and admired youthful heroes. Members stared and whispered among themselves as he strolled into the smoking-room of his London club. He was applauded when recognised in the streets. He visited the auditorium of schools where the head boys greeted him by calling for 'three cheers'. He opened church bazaars and signed masses of autograph albums. A fashionable preacher named Evelyn Wood as a perfect example of 'Christian manliness'. Wood accepted the hero-worship with becoming modesty and charm. But he was always glad to get away from it all to follow the greatest interest and passion of his life next to the army— the hunting and slaughtering of wild animals. Hunting, like sex and eating, is a human appetite becoming in moderation, but unlovely in excess. In Wood's two-volume autobiography *From Midshipman to Field Marshal* his oft-recurring hunting reminiscences flow through its chapters like a polluted and stinking stream. These become a self-indictment of a 'Christian manliness' which brought so much terror and pain to God's lower creatures with callous indifference to their sensitivity and beauty.

In India he did much pursuing of tigers, as well as of rebel natives, and he wrote boring accounts of such activities. Any living thing in a state of nature was his target. He wanted to impress some natives with his rifle and his aim, so he shot the head off a squirrel at a distance. He adored pig-sticking, and he had a thrill when three hunting companions threw stones at an exhausted boar. The boar charged Wood, but a spear from a companion saved him. Proudly he recorded that on this pig-hunt they attained 'what must be a record—five pigs in an hour and a half amongst four sportsmen'. Another day, after 'a long, fruitless hunt for tigers' they found on the way home a bear 'which fell riddled with seven bullets.' 'I was much impressed by the excitement of the hitherto placid elephant, which trumpeting loudly, knelt violently on the bear, crushing it flat and then tossed it between fore and hind feet, as if playing with a ball.' But, of course, hunting was a cherished pastime with the officers of the armies of Victoria. Wood related how, when officers arriving in the Crimea to fight the Russians learned it was good hunting country, they sent back to England for hounds.

Tipped as 'an officer to watch', Wood became a student at Staff College on his return to England. There he worked with zest, nevertheless finding time to hunt foxes five days a fortnight and to take lessons from professional pugilists in London. In the summer of 1866 he started his first long sojourn in Aldershot. First he was briefly—but much to his distaste—Deputy Assistant Quartermaster General in charge of the Instructional Kitchen of Cookery. He could not draw but he had to hold the cookery post together with that of Instructional Officer in Military Drawing and Field Sketching. After that he was decently appointed Brigade Major in North Camp. The Aldershot G.O.C. was at that time the Hon. James Yorke Scarlett, second son of the first Lord Abinger. A soldier of great private wealth, Scarlett was respected by Wood as 'a gentleman in the highest sense of the word'. Scarlett surpassed his troopers in his capacity for swearing and blasphemy, but his gentlemanly qualities were revealed in his extraordinarily chivalrous attitude towards women, regardless of their social class. For example, Scarlett felt it undignified that mothers leaving hospital after childbirth should have to walk back to married quarters with their babies or be transported in an army wagon. Thus he ordered that every mother and baby should leave hospital in a hackney carriage and the bill be sent to him. Scarlett was also known to have paid for 'decent funerals' for dead children of soldiers' wives who lived 'off the strength' in Aldershot's slummy West End. In his time local undertakers advertised their services for the burial of children with 'a neat coffin, carriage, pall, etc. from fifteen shillings, with no extra charge within five miles'. Wood and Scarlett shared a bond in both having been veterans of the Crimean War. While Wood was the humble but belligerent midshipman, Scarlett was gallantly commanding the Heavy Cavalry Brigade at Balaclava.

In 1867 Wood had married Paulina, sister of wealthy Viscount Southwell. She was a Roman Catholic and he a Protestant. But what disturbed her Roman Catholic family was that Wood, by its standards, was quite poor. The V.C. who had fearlessly challenged and overcome cut-throat native rebels and bandits, shrank from a personal confrontation with Paulina in proposing marriage. He made his offer through the post. Predictions of the Southwell family that divergencies in the faiths of the couple would lead to marital strife never materialised. Wood remained a staunch member of the Church of England of the evangelical school, although he was once rebuked by the Secretary for War for permitting the intoning of prayers at Aldershot church parades under his supervision, such intoning being against army religious regulations.

For three years from November 1868 Wood was Deputy Assistant Quartermaster General at Aldershot. During that period he found time to enter as a law student in the Middle Temple and almost simultaneously to enjoy sport with a pack of drag hounds which he purchased and maintained near Rivenhall in Essex. Without difficulty he passed his law examinations and became a member of the Bar. After varied service in the British Isles he went off with Sir Garnet Wolseley to the Ashanti War in 1873. In West Africa he raised a regiment and was slightly wounded in the battle of Amoaful. He was mentioned in despatches five times and on his return home was made a Companion of the Bath. Again he was being idolised by the nation and again he was being held up as a typical example of Christian manliness.

Aldershot welcomed back the hero as Superintending Officer of Garrison Instruction. His duties included lecturing to groups of Sandhurst cadets who had been dispersed among several army stations when an outbreak of scarlatina closed their college. A complaint made to Wood by a Sandhurst instructor exemplifies the farcical situation then existing in the training of officers. The instructor told Wood: 'Although I have got as gentlemanlike a set of young men under instruction as it is possible to find in the whole world, it has been heartbreaking to try and keep them together for concerted work. First of all, their mammas and their sisters wanted them to dance all night, while by day they were constantly away at Epsom, Ascot and Goodwood. Then I hoped, the season being over, I should get them to work, but with the middle of August came requests for leave for grouse-shooting, followed in September by applications for a few days' partridge-shooting, and early in October outlying pheasants demanded attention. Now one of the best young fellows in my class wants leave for cub-hunting.' 'Oh,' replied Wood, when he heard this, 'you should put your foot down. Tell him to cub-hunt at daylight, and get here at ten o'clock.' 'He would do that cheerfully,' answered the instructor, 'but he is Master of the Hounds, and his kennels are 200 miles from London.' At this time Wood was, of course, regularly dissipating many hours of his time in hunting.

Next Wood was Assistant Quartermaster General in Aldershot. Then he was appointed to the command at Chatham, only to be called away very soon afterwards for the Kaffir and Zulu Wars. Absent for three years, he was again daring and brave. He was mentioned in despatches fourteen times. Once his horse was shot from under him. He thought that now he would remain for some time back in command at Chatham. Duty intervened, however, and called him back to Africa as deputy to General Sir George Pomeroy in efforts to quell a Boer revolt. Pomeroy was killed and his men defeated at Majuba Hill. While the public at home clamoured for revenge, Wood warned the government that it would take several weeks before he could start to retrieve the situation. Then Prime Minister Gladstone and his cabinet decided that, instead of fighting, Wood must negotiate and reach a peaceful settlement with the Boers. With stiff upper lip Wood, who would have preferred to fight, carried out the cabinet's instructions while a strong faction in England attacked him for 'a disgraceful surrender'. Wood lingered on in Africa as a royal commissioner for settlement of the Transvaal. He came back to Chatham with promotion to major-general, and with more glitter for his tunic in the form of the Grand Cross of the Most Distinguished Order of St. Michael and St. George.

He also brought back from Africa lustre to his hunting prowess to be cherished in his own memory and to be regaled to others. There was, for example, that unforgettable day twenty-five miles outside Pretoria when he and his aide-de-camp 'enjoyed some good runs after a herd of wild ostriches'. 'We chased them with hunting whips merely for the pleasure of a gallop, for when the birds could run no farther we left them to recover their breath.' But this was in 1881 when an officer could enjoy such fun and still be regarded as a 'Christian gentleman' rather than a heartless cad.

In 1882 Wood was with Sir Garnet Wolseley in Egypt for the quelling of the uprising of Arabi Pasha. Late that year he was appointed sirdar (commander-in-chief) of the Egyptian army. Since, in the rebellion, the army had been disbanded, Wood had virtually to re-create it, a task he accomplished with efficiency and speed. Then, when the expedition was organised for the rescue of Gordon in Khartoum, Wood was in command of the lines of communication. Despite the failure of the expedition, Wood added to his fame, his prestige and his decorations.

Wood's reputation as a 'daring' reformer began when he was given command of the Eastern District in 1887, with headquarters at Colchester. Under the stubborn conservative hand of George Duke of Cambridge, the British army still wasted most of its time marching and drilling to no purpose, and to guarding all kinds of places without reason. Mass marching was, more than anything else, a spectacle of turning and wheeling and changing formations in complicated manœuvres.

It was no training for war. At Colchester, Wood initiated day
and night marches and also long-distance cavalry rides under
strenuous conditions. He made troops find their way not by
maps, but by compass. He impressed upon them that wars
were not fought on barrack squares or parade grounds but
often in wild, unsurveyed country. He derided the antiquated
movements of traditional drilling procedures, asking, 'How
can these help you to defeat the enemy?'

In his autobiography Wood recalled how the Empress
Catherine of Russia posted a sentry over a plant which she
specially valued. Soon afterwards the plant died, but a sentry
post, with soldier perpetually on guard, was maintained on the
spot for 100 years. Wood commented that he found various
sentry posts at Colchester to be equally redundant and absurd.
'Go away,' he ordered a sentry whose duty was to 'guard' his
office. 'What folly!' he exclaimed on spotting a sentry outside
the 'prisoners' ward' of a garrison hospital which had not
housed a prisoner for three months. Indeed, many sentries
who by tradition guarded nobody and nothing were ordered
by Wood to return to their units for more useful duties.

Soon afterwards Wood focussed his attention on the soldiers'
canteen which the men used only when there was no other
possible alternative. They would ignore the canteen for
recreation and fun in nearby public houses. In the canteen the
beer was usually bad. The food was stale, unattractive. These
had to be consumed from stained and greasy barrack tables by
the customers who sat on shaky wooden forms. Wood arranged
for the men to be supplied with decent beer by a reputable
local brewer. He replaced the ancient barrack tables with
tables with marble tops, at that time regarded as elegant and
fashionable. He threw away the wooden forms and introduced
armchairs, despite warnings by officers that the men would soon
smash them up. Finally, Wood added to the glamour by
having mirrors hung on the canteen walls. In such daring
improvements he again risked the disapprobation of the Duke

of Cambridge who, while enjoying his reputation as 'the soldier's friend', frowned on reforms that might 'pet and pamper' them. No wonder that the duke opposed suggestions that Wood should succeed one-armed Lieutenant General Sir Archibald Alison* as Aldershot G.O.C. But the duke was overruled by the Secretary of State the Hon. Edward Stanhope. Wood moved to the Aldershot command in 1889.

The new G.O.C. rejoiced that he found, awaiting his pleasure, what he described as 'a staff of polished gentlemen'. Into his personal entourage was invited Captain Hew D. Fanshawe, a highly polished gentleman who, to the reasoning of Wood, was a capable officer because he was so good at hunting and shooting game. 'In spite of being badly off,' recorded Wood, 'Fanshawe hunted the regimental pack of staghounds while quartered at Norwich.' And 'he was known to have got single-handed, in Arran, on 12th August 1888, 161 grouse with 200 cartridges'. Moreover, 'Fanshawe's horses never refused their food, though I have known him rail them to Reading, ride seventeen miles to meet the South Oxfordshire hounds, and return at night to Aldershot. Such practical sportsmen were of great assistance to me in the outdoor work which now engaged most of my time.' While admiring the dash and panache of Fanshawe and all other polished gentlemen of the Wood era in Aldershot, it is hard not to associate their spirit and outlook, lingering until the time when Douglas Haig was G.O.C., as contributing much towards the British army's strategic débâcles in the First World War. Wood was obsessed by the belief that the man who excelled on the hunting field would also excel in the field of battle. Once he made the fatuous claim that the possession of 'a good seat', meaning the manipulation of the bottom on the back of a horse, was of more value to an officer than 'a good head'.

However, under Wood and his staff The Camp did make considerable progress. Living conditions of the troops were

* He lost an arm at the siege of Lucknow.

substantially improved. When Wood became G.O.C. the
majority of the men were still living in draughty, leaking huts
built in 1855 and units who occupied existing stone and brick
barracks were greatly envied and were in health immeasurably
fitter. Wood advocated replacement of the huts in the North
and South Camps by barracks built in company blocks. Then
he wanted each barracks to be given the name of some outstanding British victory. After fighting many objections, Wood
won sanction for the great reconstruction plan, the cost of
which exceeded £1,500,000, then a tremendous sum to be
spent for the accommodation of mere privates and N.C.O.s.
Progress in the building of the new barracks was frequently
interrupted by labour troubles. There were complaints, even
in the 1880s, that the British worker had become so pampered
and spoiled that all he thought about was money. Nevertheless
building labourers at Aldershot were being, in many cases, paid
less than those who worked on The Camp in the 1850s. They
went on strike for a substantial increase in their pay of 3½d. or
3¾d. an hour and many would walk to Farnham each night to
sleep in its workhouse. And then came the day when even
skilled bricklayers were demanding more money, even though
they were receiving as much as sevenpence an hour.

As at Colchester, Wood swept away many unnecessary
sentries. One sentry-box, with a soldier on duty night and day,
was at the entrance to the hospital. Nobody could explain the
reason to the G.O.C. except to say, 'But he's always been
there.' When Wood spotted the hospital sentry he also noticed
soldiers entering the building with tin cans of food. He asked
for an explanation, and was told of a rule that food for the
soldier patients must be prepared in their regimental cookhouses, some more than half a mile away. Naturally meals
cooked at a distance and carried to the hospital were cold on
arrival at the patient's bedside. But Wood soon brought sanity
into hospital feeding arrangements. He also studied what all the
ordinary soldiers of The Camp were given to eat. On his

intervention meals were improved, but these were far below the army standards of today or even of that of the early twentieth century.

Wood was usually in debt. Indeed, he borrowed £3,000 with which to pay the expenses of moving from Colchester and establishing his household in The Camp. Nevertheless he was hard on officers who lowered the reputation of the Command by running up big bills with Aldershot tradesmen. On finding that the liabilities of the Officers' Club exceeded £1,000 he ordered every member to contribute a sum towards paying this, first handing over £100 himself. When the Leinster Regiment arrived at The Camp, Wood enthusiastically welcomed a suggestion that it should revive in Aldershot the ancient custom of 'Crying down the credit'. The regiment's drum-major and two drummers forthwith marched through the main streets of the town 'beating the taps' to warn tradesmen against giving trust.

Under Wood the officers and troops of The Camp had to work harder—not that he much disturbed the hunting habits of the former. He demanded a higher standard of marksmanship among the infantry. He revolutionised the practising drill of the artillery which, until his arrival, had slavishly followed the procedures of the Peninsular War of 1807–14. He insisted upon a big increase of night training, with mock assaults on trenches, and even cavalry skirmishes, taking place in darkness. But ceremonial drilling and marching, which could have no usefulness in war, were greatly reduced. Lord Wolseley, then Adjutant General, followed the activities of the Aldershot G.O.C. with enthusiasm. Wolseley suggested that Wood should give similar modern training to troops at other army stations. Commander-in-chief George Duke of Cambridge, however, observed Wood's Aldershot exertions with growing disfavour.

In this mood of disapproval the Duke of Cambridge arrived in Aldershot one morning to review the Cavalry Brigade and

I

then to be entertained to a luncheon of his favourite pork chops and other viands by Wood and his staff. The review went off well and the commander-in-chief seemed very satisfied with the state of the Cavalry Brigade. Then, suddenly, in the presence of a big group of important officers, the duke turned towards Wood and berated him in strong language. The duke told Wood that he had become aware of the night training operations and regarded them with the greatest dislike. The duke went on to make a special protest against the use of horses at night. Such activities, he explained, interfered with the horses' rest. While receiving the commander-in-chief's tongue-lashing, the Aldershot G.O.C. stood at attention. A subordinate officer remarked afterwards that 'not a muscle of his face moved'.

The duke, usually so cheerful at mealtimes, had another shock for Wood. He declined to attend the luncheon prepared for him and suddenly rode away with his entourage to Farnborough station to board the next train home to London. Wood, back in Government House, where the rich but rejected meal had been prepared, was filled with dismay, at the same time conscious that his commander-in-chief, who loved eating, would be ravaged by hunger as well as anger. With a sudden brainwave, Wood had a large basket filled with food. He handed this to a Captain Babington with orders that he deliver it to the commander-in-chief with greatest despatch. Babington, later described by Wood as 'the finest horseman in the Division', thereupon with basket on one arm, mounted a charger for a gallop to Farnborough station. All who saw this gallant officer, riding even faster than John Gilpin, were puzzled by the irregularity of the basket swaying on his arm. But he was able to hand this picnic lunch to the duke's aide-de-camp before the train steamed away.

News of the Aldershot incident quickly spread. It became the talk of all those London clubs where officers had membership. For the Duke of Cambridge to have rebuked General Wood in the presence of officers of lower rank was criticised as ill-mannered and unkind. Wood, bound by the conventionalities of the day, formally reported the duke's rebuke to Adjutant General Lord Wolseley, expressing willingness to resign the Aldershot Command. Wolseley, who refused to take the duke —'that great German Sausage'—seriously, requested Wood to remain at his post and 'to go on as you are doing', which meant that he might continue night training, even if it did cause horses to lose sleep. However, there were indications that the commander-in-chief regretted his boorishness. On later visits to The Camp he even praised Wood and allowed himself to be entertained to lunch.

The Duke of Cambridge made another scene at Aldershot, but for this Wood can be absolved from blame. The occasion was a lecture attended by both the duke and Lord Wolseley in the Cavalry drill hall at Aldershot in April 1890. The lecturer was a Colonel Bowdler Bell, a progressive student of warfare who might be described as a Captain Liddell Hart of his age. With the Duke of Cambridge sitting a few feet away from him, the lecturer had the temerity to hint that the days of the usefulness of cavalry were passing; that increased fire power made mounted men easy targets. He suggested changes in the army and its battle tactics. As soon as Colonel Bell had stopped speaking, the commander-in-chief arose from his seat. Looking round sternly, he said: 'My impression is that the army has done very well and should not be eager to change. An opinion has been expressed that the days of cavalry are over. My own view is exactly to the contrary—that the days of that arm have in fact come in!' The meeting soon broke up in embarrassed silence. Certainly Wood would have agreed with the commander-in-chief on this occasion and would have declared Bell's sacrilegious questioning of the usefulness of the beloved and sacred horse.

The duties of Evelyn Wood as Aldershot G.O.C. did not interfere with his passion for killing animals for fun. He frequently darted away to hunt stags in Devon, to shoot birds in Essex and to pursue foxes in various parts of the United Kingdom. Of course, he hunted locally too. He was excited to discover that hares were living in the grounds of Government House, and he proclaimed that territory to be their preserve. But he allowed these timid creatures to be hunted by the Foot Beagles, thus (he wrote) 'affording officers and men much amusement'. The hares were also trapped and used for coursing. Wood rendered valuable service to the local Mr. Garth's Fox Hounds when it was reported that the area had been so denuded of foxes that 'kills' were difficult to obtain. He persuaded Lord Cork to send him some families of fox cubs from his estate. Wood had a large enclosure made for them outside Government House. When they were sufficiently old and strong they burrowed themselves out of the enclosure to move into the countryside to sacrifice their lives to Mr. Garth's Fox Hounds.

In 1890 Lady Wood died. Evelyn's marriage to this self-effacing Roman Catholic had been happy. She remained as far as possible in the background as befitted the wife of a national hero. But of course she followed the traditions of the Aldershot commander's wife in visiting hospitals, arranging sewing parties and being hostess at Government House's frequent official dinners and lunches. But she preferred her family circle to the charity work and entertaining. The Woods had four devoted children, three sons and a daughter. With Paulina Wood's death the attractions that Aldershot held for Evelyn seemed to have turned into a dull routine. Thus he was glad to leave The Camp to become Quartermaster General to the Forces in the autumn of 1893.

In addition to his new exalted military appointment, Wood was also made Prime Warden of the Fishmongers' Company of which he had long been a liveryman. Told that his office entitled him to select the guests for the Company's annual banquet Wood decided they should all be fox-hunters. Edward

Prince of Wales was there, together with the Duke of Beaufort, forty-three other masters of hounds and 200 additional hunt members. Many guests at this unique banquet were bucolic in appearance and wild in their manners and behaved as if unaccustomed to formal banquets in the capital. One newspaper suggested the affair would have had more appeal and been more appropriate if the Fishmongers' Company had instead entertained the nation's anglers. In taking office there were many problems for the new Quartermaster General to solve. Nevertheless the ample leave granted to officers at the time permitted him to spend sixty days of the year in hunting and shooting. In Ireland he recorded a marvellous day of sport when beaters in an osier bed frightened out 'a deer, a fox, lots of pheasants, partridges and some wild duck'.

As Quartermaster General, Wood became involved in an unusual controversy. This became known as the Battle of the Urinals. The facts of this controversy reflect the shocking lack of consideration with which soldiers were treated less than eighty-five years ago. At the hour of 'Lights out' in British barracks they really went out and it was a grave offence to light a candle or even to strike a match. Thus when at any time after 'Lights out' a man wanted to relieve nature he had to feel his way through the darkness to the insanitary lavatory tubs placed in the corner of the room or in the dark passage outside. Because these tubs could not be seen by the unfortunate men trying to use them, the floor around became in a filthy, repellent condition. In some of the Aldershot barracks men preferred to avoid the dirt and the stench of the tubs by climbing through windows on to the balcony that ran along one side of the buildings. They urinated from this to the yard below. But there were several cases of soldiers who, half asleep or dazed by alcohol, climbing through a window on the side where there was no balcony, thereby plummeting to the ground. In 1890 a Private Mallock was killed in such a fall at Aldershot. A few months later there was a similar death. The cases caused puzzlement among the uniformed public because, at the inquests, coroner and witnesses in Victorian decorum would only say that the deceased had left the barrack room through its window 'for a certain purpose'.

In an official inquiry into the drop in recruiting, it was stated that soldiers' stories of the horror and perils of going to the lavatory by night had prevented many sensitive youths from enlisting. Therefore Wood proposed that a urinal-light system should be installed in military centres throughout the British Isles. The Accountant General was incensed at this proposal thus to spoil the troops. He estimated that the installation of urinal lights would cost £3,000 and that their maintenance would mean the expenditure of another £2,000 a year. Some army and War Office officials sided with the Accountant General in this pungent Battle of the Urinals. Finally, Wood appeared before the Army Council and, without preliminary apologies, presented a stark and repulsive word picture of an army tub and its surroundings after it had been used throughout the night in darkness. Lord Lansdowne, Secretary for War, was so appalled and repelled that he threw his powerful support on the side of the Quartermaster General. The urinal lights were duly installed.

As Quartermaster General, Wood reorganised the system of army transport and supplies. And at the age of sixty, still the intrepid sportsman, he learned to ride a bicycle. At that time to pedal about the crowded streets of London was regarded as daring, advanced, but hardly an occupation for an officer and gentleman. He was pained when one morning in the Edgware Road a horse pulling a hansom cab became alarmed by the proximity of the illustrious bicyclist and buried its teeth in his arm. One of the cycle wheels got wedged under the cab and horse until a blacksmith was engaged to file some of the metal away. On another unhappy day the general suffered a head-on collision with an omnibus horse outside the Mansion House. And there was still another disaster when a fast-moving hansom cab, in the words of Wood, 'took the cycle away from underneath me, carrying it seventy yards before the driver could pull up.' Wood liked to boast that in one twelve-month period he cycled a total of 2,000 miles. Often he pedalled the thirty-six miles between London and Aldershot.

In 1897 Wood was made Adjutant General. Within two years he was mobilising the forces for the South African War, raising and organising a succession of formations and their shipment. On the evening of January 22, 1901, the queen whom the sixty-three-year-old Wood had served for so long, died. The funeral procession on the icy morning of February 2 was one and a half hours late in starting. Then the cream-coloured horses drawing the gun-carriage which bore the coffin, refused to keep pace with the slow-marching infantry. There was anxiety at Windsor station when the horses were being attached again to the gun-carriage. In a nervous condition they threw themselves about when massed bands started to play. The gun-carriage, with its precious imperial corpse, alarmed Wood by 'rocking ominously'. The gun-carriage's swingletree broke. The horses were abandoned and sailors instead hauled the coffin to St. George's Chapel.

On ceasing to be Adjutant General in 1901 Wood began his last active command which was over the Second Army Corps, with headquarters at Salisbury. The rhythm of his life continued, with the alternations of soldiering and hunting. He appointed his son Captain C. N. Wood as his aide-de-camp. An important part of the young man's duties was to find his father lodging and stabling for his innumerable hunting excursions. His favourite location became the Blackmoor Vale Hunt country. His verdict on its population was that 'a more pleasant set of hunting gentlemen it would be impossible to imagine'. There was Mr. John Hargreaves, an old Aldershot friend, 'who carried the horn himself'. In the 1901–2 hunting season he and Hargreaves 'accounted for a 100 brace of foxes'. At the end of these eventful months Wood sped to Melbury to visit Lord Ilchester in the vicinity of whose home were 'twenty-two litters of cubs', doubtless all doomed to slaughter by the hunting gentlemen. Later, inspecting the North Devon Regiment, Wood was impressed and gratified to learn that among its officers were 'eleven Masters or former Masters of foxhounds'.

The general who as Quartermaster General had fought and won the Battle of the Urinals continued his efforts to make life sweeter and more sanitary for the troops. In the Second Army Corps District he flouted the old and respected tradition that the commander, while inspecting the front and interior of barracks, ignored the back of these buildings where the men's latrines were situated. He would arrive at barracks gates without advance notice and despatch a member of his staff to inform the senior officer that he wished to inspect the latrines immediately. Then he would proceed to these with his entourage. Conditions found on such first visitations were shocking. Barracks officers likewise considered these prying inspection of privies not only shocking but 'in damned poor form'. Nevertheless in the Second Army Corps District they achieved a welcome sanitary revolution.

Wood fought a battle over soldiers' shirts, a battle that caused resentment in the Ministry for War. Each of his soldiers was allowed two shirts. Wood decided these were insufficient. The troops wore shirts by day with their uniforms. They also kept them on in bed as nightshirts. They complained how, when one shirt was being washed, the remaining one often got wet in bad weather so could not be worn in bed. Wood hinted that the two-shirt rule was mean and stingy anyhow, whereas officials in London felt that Wood's demands were extravagant and that to accede to them would result in unwise pampering of the troops. However, Surgeon General G. J. Evatt of the Second Army Corps District prepared for Wood a document containing graphic descriptions of conditions in some of the regiments caused by such strict rationing of shirts. These were forwarded to the Minister for War who, overruling his subordinates, ordered that the shirts of the troops be increased from two to three for each man.

In April 1903 a letter reached Wood from the royal yacht *Victoria and Albert*. This was from King Edward VII telling him that 'it has given me the greatest pleasure and satisfaction to promote you to the rank of field marshal, after the long and distinguished services you have rendered for Crown and Country'. Wood retired in the December of the following year. Now sixty-six, it was fifty years previously that he had started his dedication to 'Crown and Country' as a midshipman. Wood lived on in Harlow, Essex, for another fifteen years, for most of these still participating in his beloved hunting and shooting. The man who had lashed ostriches with whips, shot heads off squirrels and had triumphantly chased foxes, stags and other creatures to their doom felt, in his latter years, a righteous hatred for all those who campaigned against such manly 'sports'. This hatred would from time to time explode into letters which he despatched to the *Morning Post*, *The Times* and other responsible national newspapers. Here is one such letter, written in all seriousness in the dark war days of 1917:

'While I do not assert that Sir David Beatty and Sir John Jellicoe won the Battle of Heligoland and the Jutland Bank by their practise of fox hunting, which certainly braces the nerves, it may interest the objector to fox hunting, whose arguments I am answering, and who is a well-known journalist and writer on naval affairs, to learn that the Vice-Admiral is said to ride brilliantly to hounds, and that as late as the season 1913–14 I saw the Admiral riding well in front of the Essex pack.'

Thus did the ancient field marshal waffle and write until, like old soldiers in the famous army ditty, he 'simply faded away'. Then came the final glory and honour of the lavish military funeral at Aldershot for one who, in the opinion of the day, had served his country well and had always 'kept a good seat'.

7

Hero's Fall

WHEN Arthur Duke of Connaught ended his term as
G.O.C. Aldershot in 1898 the announcement of his successor
caused pride and pleasure. For the new commander was General
Sir Redvers Buller, v.c., a national hero idolised by the public
and adored by the army. The tall Buller was now aged sixty
and wide of girth, giving the impression of flabbiness that
comes to elder men who live too well and do not exercise. For
he had spent nearly all the ten previous years sitting at a War
Office desk as Quartermaster General and Adjutant General.
Buller's sedentary life had not been improved by a plenitude
of rich foods and fine wines. Nevertheless in romantically blind
English eyes he had remained the dashing and intrepid hero and
leader of men. Hundreds of parents were still naming their
baby sons Redvers in the hope they would grow up with the
manly qualities of this General Sir Redvers Buller. In those
days the stars of pop, films and sport had not arrived to over-
shadow bold and brave army leaders as the heroes of British
youth. Thus the story of Buller was unapologetically related
in schools, and with blood-curdling illustrations in popular
magazines and cheap booklets.

The story of Redvers Buller was indeed remarkable. One of
eleven children, he was born into wealth and inherited the
manor of Downes, near Crediton in Devonshire. His father
James Wentworth Buller had two periods in the House of
Commons as Liberal M.P. for Devonshire constituencies. His
mother was a niece of the 12th Duke of Norfolk. Redvers had
the slightly receding chin which often distinguishes members

of the British aristocracy. His education was that of an English gentleman, except that he went to both Harrow and Eton. The boy's stay at Harrow ended in expulsion for a number of pranks which included painting doors red. His active service started in India and then in China at the age of twenty-one. Ten years later he joined the Red River expedition in Canada as a lieutenant of the 60th Rifles.

The outward manner of young Redvers was rough and blustering. He drank heavily. But Sir Garnet Wolseley (later Lord Wolseley), the commander of this 600-mile expedition from Lake Superior to Fort Garry, quickly saw through the façade and recognised the young officer as one with a promising future. Amid the wild and perilous adventures of the expedition, Redvers never failed to show initiative and courage. While his brother officers and men were uneasy and perplexed by the hard unfamiliarity of the Canadian wilds, Redvers behaved as if he were a native of them. He proved himself a superb canoeist on the swirling streams and rapids. In tracking and in his ability with the axe, Redvers surpassed the skills of Canadians who had spent all their lives in these regions. He could also outrival them in roughness of manner, in crude indecent talk and expletives. However, the sharp perceptive Wolseley was never fooled. He got to know another Redvers who privately revelled in Greek and Latin verse, and who always carried in his pocket a miniature volume of the essays of Bacon. He learned that Buller was a religious man, but with a philosophy which came more from the Old Testament than from the New. And at his roughest, Redvers possessed a personal magnetism that caused men of all kinds to follow his leadership.

In 1874 Buller was serving, like Wood, under Sir Garnet Wolseley in the Ashanti campaign in West Africa which resulted in the capture and the burning of Kumasi. Buller became a hero of this war. Stories of his exploits for queen and country against the rebellious 'savages' enchanted Victoria and she received him with delight on his return home. And now he was hailed by Wolseley as 'a veritable god of battle who draws men into a fight like a magnet does steel'. Buller had indeed a power that could at will draw all men unto him. They would follow him unhesitatingly in war exploits of extreme danger and against seemingly impossible odds. In the words of Wolseley's American biographer Joseph Lehmann, 'Buller had a kind of animal magnetism that cast a spell over those who served him'. Soldiers related how, amid the terror of battle with hordes of fanatical tribesmen, they would suddenly find calm and confidence when joined and urged forward by Buller. His was the power of a father in the presence of his young children. When Buller was near they felt safe and secure, even in the proximity of a thousand enemy spears. In the jargon of modern psychiatry, Buller was a great 'father figure'.

The 'god of battle' was back in Africa in 1878 and 1879 to become exhilarated by the bloody Kaffir and Zulu campaigns. In the latter he won the Victoria Cross for saving the lives of four men in defiance of all danger. The first—a wounded soldier—was in the middle of a wild mob of Zulus. Buller plunged into the mêlée, dragged him out and carried him to a place of safety. But the soldier was, alas, killed some hours later. Buller then rushed back towards the Zulus, who were rapidly advancing, and saved a second man in the same way. Next Buller returned to save a Captain D'Arcy and later a Lieutenant Everett. Buller had an almost pathological hatred of newspapers and their reporters. Nevertheless, despite the snubs he gave the war correspondents, they wrote dramatic stories of his prowess in battle. One described Buller as leading his men on horseback 'with the reins in his teeth, a revolver and a knobkerrie he had snatched from a Zulu in the other, his hat blown off in the mêlée, a large streak of blood across his face, caused by a splinter of rock from above, this gallant horseman seemed a demon incarnate to the flying savages, who slunk out

of his path as if he had been—and indeed they believed him—
an evil spirit whose very look was death'.

He came back to England for service at Aldershot where at
the time the G.O.C. was Sir Thomas Steele. But within a year
he was in the thick of the first Boer War of 1881 as chief of
staff to Sir Evelyn Wood. Then, in 1882, in a brief respite
between campaigns, he came home—and married. Buller was
then aged forty-three. And the marriage surprised many of his
friends. They had regarded him as a confirmed bachelor who
preferred the company of army comrades to any woman. It
was known that Redvers passionately worshipped the memory
of his mother who had died when he was aged sixteen. At
Downes he made a special garden of the flowers which he
knew she had personally planted. Wherever he travelled he was
never without her picture. It was not surprising then that his
bride was a mature, motherly woman with close associations
with his family. Lady Audrey Jane Charlotte Howard was
the widow of Redvers's cousin, the Honourable Greville
Theophilus Howard. Her father was John, 4th Marquess
Townshend. She nearly always dressed in black, as if mourning
her first husband. Redvers and Audrey were honeymooning on
the Continent when a telegram arrived from Wolseley, now
raised to the peerage, asking him to serve in the campaign in
Egypt. There had been a nationalist rising against foreign
domination, with a massacre in Alexandria. The British and
French fleets had in retaliation bombarded Alexandria. Now
Lord Wolseley was to command an expedition to crush and
punish the rebellious fellahs and their leader Ahmed Arabi.

Thus Redvers Buller abandoned love for war. He was soon
with Wolseley in Egypt as head of the army's intelligence
department, thrusting himself into bloody fighting wherever
and whenever he could find a chance. Letters that Redvers
wrote to his wife reveal that, for him, fighting brought relief
to a terrible lust, a lust for which he candidly confessed shame.
Like a virgin youth lured into a brothel, he suffered agonising

remorse after shooting and killing had apparently given him a form of orgasm, emotional or physical. Of one battle he wrote to Audrey: 'When I thought of the coming fight my spirits rose and I felt so happy, and then I thought of you and that you might be made miserable, and I felt it was really wicked to be glad there was going to be a fight. I do believe that it is very wicked and very brutal, but I can't help it: There is nothing in this world which stirs me up so much as a fight. I became just as if I were aerated. I always feel very much ashamed of myself afterwards, but I cannot help it.' Buller's lust for battle was catching. His frenzied desire would somehow be communicated to his men who would follow him with mad elation to shoot, bayonet or hack down the enemy. They too felt 'aerated' until the victory was won and was proven by the masses of slaughtered Moslem youths and men, lying on the stinking fly-blown battlefield.

Drama, adventure and battle returned to the life of Redvers Buller in 1884 when he served as second in command to Major-General Sir Gerald Graham in the campaign to subdue turbulent Suakin and relieve a beleagured garrison at Tokar. In a crisis of the campaign he saved Graham's second brigade from destruction by daring and gallant action. But since Buller was now an acting major-general, he did not throw himself bodily into the scrimmage. With his lust for fighting Buller found such celibacy hard to endure. Thus he wrote to his wife: 'I confess I envy the people who were there in the scrimmage, and would have given, for the moment, a good deal not to have been a general, and to have been able to go for the niggers, but of course that would have been *infra dig* and I had to look on. Frenzied Mohammedans rushed to death on our bayonets. The Arabs admit to the loss of 1,500 dead.'

Later in 1884 Redvers Buller was again serving under Lord Wolseley. This time he was chief of staff to Wolseley in the unsuccessful attempt to save General Charles Gordon besieged in Khartoum by the warrior-prophet Mohammed Ahmed, notorious as the Mahdi, which means Messiah. Gordon, the bible-reading whisky-swigging eccentric, had been besieged five months before a vacillating British government decided to send an expedition to save him. Gordon's refusal to carry out any orders other than his own had not endeared him to the government or to a number of figures in the army command. On the other hand he was adored by a romantic British public and he was admired by the queen. Gordon had been despatched to the Sudan as governor to extricate the harassed Egyptians whose rule over this turbulent area had broken down. He preferred to seek inspiration from the bible and to follow an incomprehensible policy of his own.

Gordon was not one of Buller's heroes. When the idea of sending an army to rescue Gordon was first discussed, Buller's comment was that 'The man isn't worth the camels.' Apparently Buller was jealous of Gordon's popularity with the British populace and of the tremendous publicity given to him in the newspapers. The press had, of course, done much in making Buller such a popular figure. Nevertheless Buller, the wealthy and proud Devonshire squire, despised journalists as being common and vulgar. He treated them with contemptuous arrogance. Once he threatened a war correspondent with a horsewhipping because he disagreed with his despatches. When the famous journalist William T. Stead wrote a courteous letter to Buller asking for an interview, he affected to be furious, referring to Stead as a 'rascal' and 'impudent scoundrel'. And when Winston Churchill became a war correspondent in South Africa Buller described him as 'a scoundrel' for a similar offence. The 'rascals' and 'impudent scoundrels' of the press continued to depict Buller as a great soldier and national hero until, after the South African War, they helped to destroy him.

On the start of the Sudan expedition there was a widespread feeling that such brilliant soldiers as Wolseley and Buller could not fail to reach Gordon before the forces of the Mahdi entered

Khartoum to force his surrender. It was also felt that there could be no bungling in the long advance with Evelyn Wood in charge of communications. Yet the great rescue mission was a failure. And in it Buller, for the first time in his career, showed hesitation. Later he was to express regret that, when called upon to drive the enemy out of Mettemeh, he had decided that such an operation would be too dangerous and too costly. Nevertheless, the popular verdict was that he had done well in face of great difficulties.

In September 1884 this futile expedition started to ascend the Nile and progress was impeded by circumstances which nobody in the command had foreseen. When General Sir Herbert Stewart was fatally wounded Buller took command of the desert column. The approach of overwhelming forces of the enemy resulted in his commanding a retreat which his admirers described as 'brilliant'. Towards the end of January 1885 the expedition crossed the desert. Advance forces reached Khartoum on January 28 only to learn that two days previously the Mahdi's dervishes had taken the city and speared Gordon to death on the steps of his palace. It was not until 1898 that General Horatio Kitchener (later Lord Kitchener) supervised an impressive revenge on Great Britain's behalf. He defeated the dervishes at Omdurman, slaughtering 9,700 of them and losing only forty-eight of his own men. As for the Mahdi, who had died from an illness soon after Gordon, Kitchener commanded that the corpse be removed from its tomb and be thrown ignominiously into the river.

Incidentally, throughout Buller's activities in Egypt and the Sudan, eyebrows were constantly being raised at the importance he placed upon food and drink. Some thought that this interfered with his concentration upon the command of his forces and upon strategy. Certainly it did seem that Buller's appetite for battle had become rivalled by his appetite for good food. References to food in letters to his wife and others were numerous and emphasised the growing importance of eating in his life. Like Sir Winston Churchill he expected the highest British and French standards of cookery, even when he was far from his civilisation's centres of *haute cuisine*. As a youth on active service he carried volumes of Greek and Latin poetry and Bacon's essays. Now he carried tons of food. In his detailed description of the Gordon relief expedition, Julian Symons writes:* 'One of the amenities of life at Wady Halfa, for the officers, was dinner with Buller. The chief of staff appreciated the pleasures of eating. Forty camels were needed to carry his baggage, and several must have been loaded with delicacies for the table, and with the champagne that was Buller's invariable accompaniment on a campaign.'

Buller's commanding personality, his tested powers of leadership and new tales of his courage, related by the journalists he despised, demanded that he be given important duties on his return from the unsuccessful Gordon rescue mission. While Wolseley became Adjutant General, Buller was made Quartermaster General and was also despatched to Ireland to pacify the seething populace of Kerry and to reorganise the police. Then England heard how well he was doing, with the result that he was appointed Under Secretary for Ireland. But he did not stay long in this post, chiefly because of his creditable objections to the cruel eviction of Irish tenants and to other horrors of England's Irish policies. In 1890 Buller succeeded Wolseley as Adjutant General, a post he held for seven years. In this sedentary office occupation, he revealed new abilities as impressive as those shown in battle. He surveyed the chaotic transport and supply services of the army and neatly reorganised them. He laid the foundations of the Royal Army Service Corps. Surprisingly proving himself to be a man of business, he tackled army extravagance and insisted upon economies that would win the approbation of all taxpayers. Buller changed the army as Adjutant General and he himself simultaneously became visibly changed. He was much fatter.

* *England's Pride* (Hamish Hamilton, 1965).

K

His complexion became puffy and much more florid. He ate enormously, and some of his colleagues complained that he drank far too much. Nevertheless, as the antique Duke of Cambridge was gradually being ejected from his post as commander-in-chief, there were plans to appoint Buller in his place. However, political changes and the wiliness of Wolseley blocked such plans. Wolseley became commander-in-chief. Buller was appointed G.O.C. at Aldershot, succeeding Prince Arthur Duke of Connaught.

Officers and men could hardly conceal their delight when Buller became installed at Government House in The Camp in 1898. The citizens of Aldershot were also delighted. Although Buller had not been in the fighting field for so long, he was still the most admired and respected general in the British army. The ordinary soldiers of The Camp were specially grateful to him because, as Adjutant General, he had done so much to improve their conditions of life. Buller had been Adjutant General during three of those eventful Aldershot years of 1889–93 when Sir Evelyn Wood, shocked by its masses of rotting and leaking wooden huts, had pressed for more new barracks of stone and brick. Buller had supported him and had persuaded the government to provide the large funds necessary for their construction. Buller, while economising elsewhere, had never begrudged money that would improve the general lot of the men in The Camp. Then officers were pleased to have Buller as their chief at Aldershot because he had the reputation for exercising his authority without hesitation or favour; and he was grateful for work well done. True that there were moments when Buller exhibited a flaming temper, but this would soon subside and he would nurse no grudges. Buller was a man's man, or rather a soldier's soldier.

His attitude towards civilians was different. He was inclined to despise them, to regard them as inferior beings compared to those who wore the queen's uniform. At times he could be very rude to civilians. Nevertheless they accepted his arrogance as the well-earned prerogative of a great and gallant soldier. Wherever he went in Aldershot town he was stared at reverently by wide-eyed, hero-worshipping youths of all classes. He was adored by working folk, many of whom hung his picture in their homes close to that of Queen Victoria. There was only one who stood higher in their affections than Redvers Buller—and that was the queen.

It seemed that Buller was painfully aware that England might soon find herself in a war far greater and more difficult than the past conflicts with those dark-skinned races whom, without fear or favour, he endowed with the name of 'savages'. He had made his martial reputation in campaigns against 'savages', but now there were dangers of war on a large scale with white Dutchmen in South Africa or with the 'civilised' nations of Europe. At Aldershot he practised with the troops the mobilisation drill of which he, Wolseley and others had made paper plans at the War Office. He tightened still closer the links between army transport and supply. He strengthened the co-operation and understanding between all the army's services. The Buller worshippers took it for granted that, in addition to being the perfect organiser, he was also a tactical genius in the field. Thus there was pained surpise when in manœuvres involving 50,000 men he was decisively defeated by a force commanded by the Duke of Connaught. This gave rise to the absurd rumour that Buller had purposely given victory to the favourite son of Queen Victoria in order not to upset her. In his endeavours to increase general efficiency at Aldershot, Buller did not even forget to examine religious worship and the chaplains. He suspected these chaplains of slackness and lack of zeal. Therefore he would often ignore the two big churches traditionally attended by the G.O.C. to worship at a modest wooden church in the Marlborough Lines. He felt that in such a place chaplains might lapse into slackness unless his vigilant presence kept them on their toes. Younger clergymen would become nervous as they tried to lead worship of the Lord of Hosts while the rubicund general, described by Wolseley as 'The veritable God of Battle', watched them with his small, steely eyes.

In the middle of June 1899 Buller was called to London for a secret meeting with Lord Lansdowne, Minister for War. Lansdowne, without any preliminaries, asked the general whether he would undertake chief command of the troops in South Africa if, as was expected, war broke out there. Lansdowne had expected him to assent with enthusiasm, but surprisingly he demurred. Buller suggested that the command be offered to his old friend and patron Wolseley and that he himself should serve as Wolseley's chief of staff. However, when pressed by Lansdowne, Buller replied: 'Well, no man should decline to try what he is worth. I'll do my best.' Buller's reticence made a poor impression upon Lord Lansdowne. It laid the foundations for the disharmony which was to follow between these two men. Buller may have asked to serve below Wolseley as an act of loyalty to his friend. On the other hand, Buller may have suddenly realised that at sixty, overweight and long saturated with excessive food and alcohol, he was not the same person whose courage and initiative had so stirred the imagination of the nation in the Ashanti and Zulu wars. The lust to fight and kill may have become atrophied by high living and by ten years of work sitting on his bottom at an office desk. Indeed, the ever increasing self-indulgence of Buller when at leisure shocked his fellow officers. Colonel Ian Hamilton (later General Sir Ian Hamilton) described how Buller at a house party 'gobbled up all sorts of things'. Remarked an observant private: 'No one loves dinner better than Buller.' Then it seems that in real warfare Buller had never commanded more than 2,000 men. Stories of his strategic skill had been absurdly exaggerated and distorted.

Within a year of his appointment to Aldershot Redvers Buller was called away to lead the British forces in South Africa. On October 11, 1899, the Boers invaded Natal. Now Great Britain was officially at war with the Transvaal Republic and Orange Free State, who were led by iron-willed Paul Kruger, inveterate bible reader and fanatical Boer nationalist. The war did not come without warning. For several weeks in Aldershot there were preparations that culminated in a mobilisation order on October 7. Indeed, two weeks before official mobilisation the Northumberland Fusiliers had left The Camp for South Africa accompanied by a staff of army telegraphists. Several men of the Army Service Corps had departed too, and under harrowing circumstances according to a local newspaper which reported: 'There are many scenes one would well wish never to have seen, women in hysterical bursts of tears or worse, with set despairing faces and tearless eyes; half frightened little boys and girls, sobbing and looking with wondering eyes at their fathers.' The newspaper added that when the men's train steamed away, a band struck up, 'The Girl I Left Behind Me'.

News of the appointment of Buller as commander in South Africa was welcomed with enthusiasm by his adoring public and by the rank and file of the army. They felt that with this dashing general at their head, the British forces would quickly overwhelm and rout the defiant Dutchmen of South Africa. Newspapers, in extolling Buller, predicted an early and dire fate for President Paul Kruger. Men leaving for South Africa were given new strength and confidence by the realisation that Buller would be among them. On active service he was a hard disciplinarian, yet always he gave the common soldier a feeling of strength and security. In the words of journalist Archibald Forbes: 'Buller ruled his men with a rod of iron, yet while they feared him they had a dog-like love for him.'

As at the start of many other painful, fateful struggles, the British people greeted the arrival of the Boer War in an atmosphere of whoopie, carnival, crude melodrama and heroics. Writers of sentimental and patriotic songs became inspired. Preachers became eloquent about 'fighting for the right'. Frantic war fever raged in London and other big cities of the United Kingdom, but in smaller, compact Aldershot it could be observed in overwhelming concentration because this was the nation's biggest military centre. Indeed, seekers after patriotic and emotional thrills would entrain in their hundreds from London in order to participate in the Aldershot hysteria. Day after day troops marched towards the railway station from The Camp to the music of their bands, with the civilian mobs to cheer them off to war. Suffering wives, children and parents of the soldiers, there to bid them farewell, were often swept aside by the crowds who were present only for the fun and the sensation. Such slogans as 'Kruger, beware' and 'Aldershot to Pretoria' were merrily chalked on the troop trains. Usually the trains left the station to the music of 'Auld Lang Syne'.

As thousands of men moved out of Aldershot for active service overseas, thousands more moved in for intensive training. And the traditional brightly coloured uniforms were laid aside for jackets and breeches of khaki drill and for khaki topees, which were an early development in the art of camouflage. By night the newcomers filled the bars and the theatres, joining in the choruses of such songs as 'Good-bye Dolly I Must Leave You' and 'The Soldiers of the Queen'. In its patriotism Aldershot's Theatre Royal initiated what it announced as 'the new fashionable military nights' when army bands were engaged to play on the stage 'with limelight and scenic effects'. The euphoric side of patriotic emotion was, however, too often disturbed by appeals for donations for the wives and families of reservists who had been recalled from civilian life to the colours, and for those of other regulars who were already suffering privation. Buller's wife Lady Audrey joined in these appeals and helped to organise collections in Guildford and Farnham 'for the families of the soldiers of the queen', as the charity letters described them. Certainly these needed charity. For example, the government of Great Britain, then the richest nation in the world, announced the payment of the following separation allowances for families of reservists called to the colours: 'Wife, eightpence a day. Each child (girl under sixteen or boy under fourteen) twopence a day.' Even if the purchasing power of the penny was then four or five times what it is today, such allowances were ridiculously and cruelly inadequate.

Redvers Buller saw only a little of Aldershot's war fever. For three days after the outbreak of the war he left with his staff for South Africa. He was seen off on a special boat train from London to Southampton by many illustrious well-wishers including the Prince of Wales and the doddering Duke of Cambridge. His drawing-room on the train had been hurriedly panelled with yellow silk. During the journey a rich and abundant meal was served to the commander and his staff. They sailed in the liner *Dunotter Castle,* reaching Cape Town after a rough passage of sixteen days. On landing, Buller was hailed by cheering citizens who saw him as the saviour who would deliver their rich and pleasant land from the evils of Kruger and his Boers. His very presence seemed to convey a certain assurance that all would be well. As the crowds surged around their 'saviour', he neither smiled nor made any gesture of appreciation. While listening to an address of welcome he stood motionless and expressionless, causing one newspaper to compare him to a massive bronze statue. Adding to Buller's appearance of strength and power was the fact that folds of fat now concealed his receding chin. His florid complexion, tanned by sea air and sunshine, made him look healthy and younger than his sixty years.

Tragically, however, the Redvers Buller of the South African War was made not of bronze, but of plaster. The powers of strong leadership and of lightning decision had gone, together with the lust to kill, to risk everything in battle. Behind the still impressive exterior were wavering weakness, hesitation, an inclination for safety first. But still worse, he had developed what might be noble in the pacifist but what was fatal in the soldier—a fatherly compassion for the thousands of men he commanded. He had always been a tough disciplinarian but nevertheless he had loved his men as deeply as they loved him. But now his love had become reinforced by an extreme protectiveness which in warfare was an absurdity.

As British forces poured into South Africa, it was expected that Buller would swiftly retrieve the forces already beleagured there from dire disaster. On arrival he surveyed the situation. To those close to him he secretly expressed extreme pessimism, only later to cheer up and to make optimistic forecasts. The mind of this officer whose thinking had once been consistently positive, now tossed on waves of alternating optimism and pessimism. Of course he had reasons for this agonising uncertainty. The civilian armies of Kruger, many fighting in their Dutch civilian frock-coats, were putting Great Britain to shame and all Europe was laughing at her failures. Buller decided not to direct his forces in far-flung and complicated operations from Cape Town. Instead he decided that he could manage all this while on the move with a column of troops towards Ladysmith. At the same time he ordered Major-General Lord Methuen to relieve Kimberley and Major-General William Gatacre to advance towards Colesberg. Optimists in both London and Cape Town had predicted that by Christmas the brilliant Buller and his divisional commanders would have obtained mastery over the Boers. But during Advent the British forces met with repeated disasters. Nearly a quarter of Gatacre's troops surrendered to the Boers at Stormberg. Methuen was defeated at Magersfontein with heavy losses. The tragi-comedy of the commander-in-chief himself before Ladysmith caused horror in the British cabinet and War Office who conspired to hide the truth about Buller's South African follies from the nation.

Buller's first defeat in his efforts to relieve Ladysmith, where sixty-four-year-old Lieutenant-General Sir George White was in command, was at Colenso near the Tugela River. There he ordered his men to make a direct attack upon the Boers. He gobbled sandwiches as he watched the action while shells burst around him, one killing the staff surgeon at his side. Buller was himself severely bruised by fragments of shell—but he just continued observing and eating. As more and more of his officers and men fell dead or wounded, Buller's resolve was strangled by pity. He ordered his troops to abandon their assault and to withdraw. Even efforts to save the British guns were given up, and eight were left in possession of the enemy. Earlier Lieutenant Fred Roberts, only son of Field Marshal Lord Roberts, had been fatally wounded in a gallant gallop towards some British guns under withering Boer cannon-fire. British casualties in this futile engagement totalled 1,100.

Now Buller was overwhelmed by defeatism. He sat down and wrote messages that astounded Lord Lansdowne in London and Sir George White in Ladysmith. These, of course, were not disclosed to Buller's hero-worshipping public who still saw him as the intrepid, undaunted hero. To Lansdowne he cabled: 'A serious question is raised by my failure today. I do not consider I am strong enough to relieve Ladysmith. I consider I ought to let Ladysmith go, and occupy a good position for the defence of South Natal and so let time help us. But I feel I ought to consult you on such a step.' To White he wired that if he was to get his forces over the River Tugela 'one full month' of preparation would be required. He asked White how long he could hold out and then added: 'After which I suggest your firing away as much ammunition as you can and making the best terms you can. I can remain here if you have an alternative suggestion, but unaided I cannot break in.'

Buller's message to Lansdowne, which so shocked White-hall, resulted in the following reply from Lansdowne: 'The abandonment of White's force and its consequent surrender is regarded by the government as a national disaster of the greatest magnitude. We would urge you to devise another attempt to carry out relief, not necessarily via Colenso, making use of the additional men arriving if you think fit.' And from Ladysmith the horrified White informed Buller that 'I can make food last for much longer than a month, and will not think of making terms unless I am forced to. The loss of 12,000 men would be a heavy blow to England. We must not yet think of it.'

High officers in London who had known and admired the old Buller were now agitating that something be done about the new Buller, master of hesitation and defeatism. Lord Roberts, now mourning the death of his son, begged that he should supersede Buller and save a situation in South Africa which had become desperate. Roberts had already told Lansdowne that 'he [Buller] seems to be overwhelmed by the magnitude of the task imposed upon him, and I confess that the tone of some of his telegrams causes me considerable alarm. From the day he landed in Cape Town he seemed to take a pessimistic view of our position, and when a commander allows himself to entertain evil forebodings, the effect is inevitably felt throughout the army.' Buller had long been Lord Wolseley's protégé, but now Wolseley admitted his 'thorough disappointment' in him. For 'he has not shown any of the characteristics I had attributed to him; no military genius, no firmness, not even obstinacy which I thought he possessed when I discovered him. He seems dazed and dumbfounded'.

Redvers Buller was disgraced in the eyes of the few who knew of his messages inferring that Ladysmith should be surrendered to the enemy. But to most of the army and to the masses of Britain he was still the hero, even if things did go badly at Colenso. The cabinet decided that Lord Roberts should be sent to South Africa as soon as possible as supreme commander, and that he should be aided by the vigorous Lord Kitchener as chief of staff—but Buller had to be superseded with smoothness and tact. Thus Lord Lansdowne broke the news to him in these telegraphed words: 'The prosecution of the campaign in Natal is being carried out under quite unexpected difficulties, and in the opinion of H.M. Government it will require your presence and whole attention. It has been decided under these circumstances to appoint Field Marshal Lord Roberts as commander-in-chief in South Africa, his chief of staff being Lord Kitchener.' Lansdowne also remarked by cable that 'this decision may, I fear, be distasteful to you'. Nevertheless Buller swallowed the bitter pill and replied: 'I trust that any decision intended for the interests of the empire will always be acceptable to me.' Lord Roberts, aged sixty-seven, bearing the tragedy of his son's death with admirable stoicism, sailed from England on December 23 and reached Cape Town on January 10.

Muddle, indecision and failure were repeated by Buller in January in the engagement of Spion Kop. His troops reached the summit of this strategic hill without losses. In the mist and darkness they dug themselves in. But when daylight came it was realised that the trenches had been dug in such unsuitable areas that they invited Boer fire from three directions. No proper survey of the territory had been made in advance. There was a breakdown in the army's communications with the hill. Buller and staff disputed over the proper way of holding the hill. Buller waffled and wavered. Then came the retreat. Nine hundred officers and men were killed or wounded at Spion Kop. Three hundred more were taken prisoner. Major-General Sir Neville Lyttelton, one of the commanders, wrote of his chief after the débâcle: 'I have lost all confidence in Buller as a general, and I am sure he has himself.' In another futile battle at Vaal Krantz, Buller's force of 20,000 men were deftly outmanœuvred by 5,000 Boers. Roberts was by then active as commander-in-chief, and after Vaal Krantz Buller wrote to him: 'It is right you should know that, in my opinion, the fate of Ladysmith is only a matter of days, unless I am considerably reinforced.'

Roberts and Kitchener, without any great genius or brilliance, retrieved a situation which had seemed hopeless, and placed Britain on the slow, hard road to victory in South Africa. They stopped the strategy of futile frontal attacks on the Boers in their trenches and rifle pits, and instead harried them on their flanks. Buller finally relieved Ladysmith on February 28, 1900, at the cost of 307 killed, 1,862 wounded and 90 missing. Compassion for his men and almost morbid caution prevented him from making the most of his victory. Many of the troops wanted to follow in pursuit of the fleeing Boers. Buller stopped them. He did not want to risk further lives. He insisted that his soldiers deserved and would be given a good rest.

Later Buller went on to clear all Natal of the Boers. Military historians do not regard this as a great achievement since by then British power and enterprise were straining Boer resources. Throughout this Natal campaign Buller's physical appetite never flagged. Meals from his travelling kitchen were still described as excellent. Officers invited to dine with their commander particularly appreciated the Veuve Clicquot champagne which was his constant travelling companion. Once he told friends that although he preferred the best of everything, he could tolerate bread and cheese—'but not inferior champagne'. In October 1900 Buller was ordered to hand over his command and return home. This order was no disgrace, since at the time it was believed that the war was virtually over. Lord Roberts was also preparing to leave South Africa, passing the supreme command there to Lord Kitchener. Actually the Boers, with admirable gallantry and will-power, held out for another eighteen months before being starved into surrender.

Officers who had served on White's staff in Ladysmith had gossiped about Buller's defeatism during the siege. Stories of this defeatism had also leaked from the British cabinet to be whispered about in fashionable London dinner parties. Moreover Lord Lansdowne had made public a carefully censored version of despatches concerning the Spion Kop battle. These contained some of Roberts's criticisms of Buller. And next, more journalists were daring to suggest that Buller was a disappointing war commander. The nation and army at large, however, still saw him as indubitably the greatest military figure of the war, even surpassing Roberts and Kitchener who had been placed above him. Before Buller sailed for England the city of Durban gave him a bumper banquet and Cape Town gave him a bejewelled sword of honour. And when Buller reached Southampton on November 9 he received a tumultuous reception together with the freedom of that town. Even army commander-in-chief Lord Wolseley, who had privately criticised Buller with such bitterness, was on the dock to meet him, to give his hand a long clasp and to encourage his treatment as the all-conquering hero.

Buller was reappointed G.O.C. at Aldershot. The town responded by organising what for years was cherished in its history of emotional displays as 'Buller's Day'. For from the glory of the Southampton welcome Buller proceeded on the following day for still greater glories in Aldershot. This welcome home had been prepared with much publicity and expense. Early on this Friday, November 10, a series of special excursion trains from Waterloo had brought thousands of cockneys to join in the fun. As Buller's train steamed out of Southampton a hooter sounded in Aldershot so that all might know the great man was on his way. Hawkers sold 'Buller buttons' which were red, white and blue rosettes with a picture of the general as their centre-piece. At the station awaited a guard of honour. Soon there were whispers to its officer, together with an indignant little conference among the local councillors in their frock-coats and top hats. An order had been received from London that Buller must not be given a guard of honour. There were contradictory explanations for this ban. One was that a general was not entitled to a guard of honour in his own military district; another was that as Sir Redvers was officially on leave, it would be improper to give him a formal military reception. But the real reason for this sudden order seems to have been that Wolseley and the British cabinet had, following the fulsome Southampton welcome, suddenly decided that less fuss should be made of Buller. The guard of honour marched grimly away.

Anyhow, all the Aldershot council were there, and wagonettes filled with wounded soldiers, and the Bishop of Winchester, with his wife Mrs. Randall Davidson, and the children of the Wesleyan school all set to burst into 'Home Sweet Home' when Buller drove away from the station. And many 'rascals' of the press, so despised by the general, were there when, amid loud cheers, his train drew alongside the platform. One reporter, describing what he saw, wrote: 'A plump elderly gentleman, with a determined but good-humoured face, attired in a soft brown wide-awake and fawn-coloured overcoat. By his side a beaming little lady smiling through her happy tears—Lady Audrey Buller, with their daughter Miss Buller.' A local newspaper described the general as 'the doyen of Aldershot' and admitting the existence of underground criticism of his record in South Africa, commented: 'It is all very well for stay-at-homes in their feather beds to criticise, but most people are wise after the event.'

Certainly Buller was happy to be back and far away from that war-torn country where tactical decisions had been so hard to make and where soldiers were still suffering and dying. He threw himself into his Aldershot duties with zest. Large numbers of young men were enlisting in the yeomanry. Many of these were being sent to Aldershot to be formed into military units, to be fitted with clothing, equipped with arms and to undergo fast but strenuous training. By day and night such volunteers would arrive at The Camp and be given immediate attention. A friend visiting Buller one day found that he had but three officers and two sergeants to equip and organise a motley assortment of 6,000 men. Yet three weeks later he heard from Buller that more than half of these had already sailed for South Africa. Buller, in his few leisure hours, showed obvious pleasure at the popular esteem he still attracted. In Aldershot and other towns he was the guest of honour at civic and military banquets where, toying with his wine or brandy glass, he would listen to long eulogistic speeches. His solemn countenance would relax into smiles when assemblies raised glasses to toast him. And in the streets of Aldershot people treated him with even more reverence than they had done before the South African War. One cynic compared this adoration from townsfolk to that given to the Host in religious processions in Italy.

Dark clouds, however, were gathering over Buller. The press showed itself increasingly disinclined to genuflect before him. Instead, articles were printed repeating garbled reports, gossip and innuendoes regarding the defeatist messages sent by Buller to Sir George White in Ladysmith and to Lord Wolseley and Lord Lansdowne in London. Buller's position became worse when energetic St. John Brodrick (later the Earl of Midleton) succeeded Lord Lansdowne as Minister for War. Brodrick formulated a scheme for the home army to be reorganised into three corps, the first to be based at Aldershot under Buller's command. In this scheme it was explained that in the event of war, the commander who had trained his corps at home would accompany it on active service abroad. Swiftly an increasingly hostile press argued that Buller would be unfit for another overseas campaign. Bluntly newspapers accused him of having tried to persuade White to surrender Ladysmith, and to have failed miserably at Colenso and Spion Kop. Buller asked the government for permission to publish the full texts of his wartime messages to White and to Wolseley. The government refused this, offering only its own selection of messages which, complained Buller, would place him in an unfair light. The newspaper attacks grew fiercer. Now *The Times* and several other publications had opened their correspondence columns to letters denigrating the general.

Buller committed professional suicide at a public luncheon of the Queen's Westminster Volunteers in London on October 10, 1901. At this, the man who had previously treated the press and its views with disdain, issued an absurd challenge to *The Times*. Presiding at the luncheon, where wines and spirits flowed plenteously, was Sir Howard Vincent, M.P., Colonel Commandant of the Volunteers and a former official of Scotland Yard. Sir Howard proposed a joint toast to the Queen's Westminster Volunteers and to Buller. The general, his complexion looking even more florid than usual, rose and started a rambling speech defending some man whom, he said, had been unfairly criticised. Then Buller told a fantastic story about a visit paid to him at Aldershot by 'an international spy' who warned him of enemies 'who mean to get you out of the way, and they will get you out of the way, and you had better get out of it quietly'. Buller added that he had replied to the spy: 'I am a fighting man, and if what you say were really so, I am much more likely to stop here than to leave.' Next Buller said, with raised voice: '*The Times* has said I am not fit to command the First Army Corps. I assert there is no one in England junior to me who is as fit as I am. I challenge *The Times* to say who is the man they have in their eye more fit than I am.'

Friends of Buller at the table longed for him to sit down and for this embarrassing luncheon to be terminated. But next Buller was giving a muddled version of the passage about surrender in his telegram to Sir George White. He shouted: 'I challenge *The Times* to bring their scribe "Reformer" into the ring. Let us know who he is, by what right he writes, what his name is, what his authority is. Let him publish his telegram. Then I will publish a certified copy of the telegram I sent and the public shall judge me. I am perfectly ready to be judged.'

Buller returned to Aldershot to a brief but uneasy period of peace. The peace was ended when Brodrick returned to London from a country holiday and ordered Buller into his presence. There was an angry interview during which Buller made a spirited defence of his conduct and of his South African war record. The door opened and Lord Roberts, who had succeeded Wolseley as army commander-in-chief, entered. There was an altercation between Buller and Roberts, with Brodrick striving to restore calm. The two men told Buller that by making his controversial speech he had, as a serving officer, broken strict military regulations and must therefore be asked to resign. Buller refused to do this. Indeed, he appealed to King Edward VII begging that he should not be driven out of the army he had served for so long. But the king, who had read with shock and displeasure a verbatim report of the Buller speech, refused to intervene. Soon afterwards it was announced that Buller had been relieved of his command and placed on half-pay. The decision that he should go, it was reported, was made only after the cabinet had given 'long and agonising' consideration to his case.

The banishment of Buller created thunderclaps of furious public disapproval, not only in Aldershot but all over the country. In The Camp thousands of men sank into a mood of resentful protest, for they looked upon their G.O.C. as if he were father and friend of them all. Aldershot townsfolk crowded towards Government House, trying to convey their affectionate sympathy to Sir Redvers and Lady Audrey. Letters reached Buller by the sackload. These, it was reported, came mostly from working folk and many relatives of soldiers. They deplored the allegedly unfair and ungrateful treatment he had received from the government, and they assured him of their continued confidence and affection. The angriest people of all in this controversy were those of Devon and Cornwall where Buller was regarded as the greatest living West Countryman. The mayors of Exeter and Falmouth organised protest meetings. Protesters roved the countryside asking people to sign petitions to the king for the restoration of Redvers Buller to the army active service list. Buller bid gracious welcome to his

L

successor General Sir Henry Hildyard as Aldershot G.O.C. and then, with Lady Audrey and daughter, left to settle at his Devonshire manor after many years of absence. When they arrived at Crediton they were welcomed by cheering crowds and given a civic reception. Public agitation in Buller's favour continued in all parts of the United Kingdom until April 17, 1902, when the despatches written by him regarding Spion Kop were published by the government in full. These proved Buller to have been incompetent, blundering, defeatist. Together with Buller's Ladysmith messages, they proved sufficient to silence most of those who had previously regarded him as misjudged and ill-treated.

Mysteriously, however, Buller continued to be enveloped in the love and devotion of many people. Age and alcohol had made him uncomely, yet he remained a figure of almost magic glamour. His presence stirred actions of passionate demonstrativeness almost wherever he went. Once, on his way to a shooting party at a country house, he was recognised at the nearby village railway station. Quickly the inhabitants gathered. They removed the horses from the waiting carriage, waved Buller inside and then dragged it up the hill to the house of his host. On another occasion Buller was walking in the city of London on his way to a business appointment. A clerk on the top of a bus spotted him and yelled: 'There's Buller!' Almost immediately the compulsorily retired general was surrounded by a cheering crowd. Eventually he was able to plead and push his way through his worshippers to escape down a side street. He was driving with his daughter when, outside St. James's Palace, traffic was held up because crowds were waiting to see a visiting foreign royalty pass. But when the people recognised Buller they forgot about the royalty, applauding and mobbing him instead. After attending a large reception in a London mansion, Buller walked down its steps to reach his carriage. Footmen of the mansion had run ahead of him to line the steps in a kind of guard of honour and then to cheer him on his way. The cheers were taken up by the coachmen of the line of waiting carriages in the road.

Redvers Buller died at Downes, his home, on a hot day in June 1908. He was buried in Crediton. A statue of him in full uniform was erected at the end of Queen Street in Exeter. But for years after his death he lived on in the hearts and memories of men who had served under him and had felt the magic of his presence. For example, Buller's biographer and friend Colonel C. H. Melville, told of a soldier who in the First World War lost the sight of both eyes. This man, who years previously had served in South Africa, was asked if there was anything he specially wanted. 'Yes,' he replied, a 'photo of my old commander, Sir Redvers Buller.' 'Though he could no longer see this photograph,' added Melville, 'he could touch it and feel its presence.'

8

The Outsider

THE first motor car to be seen in The Camp spluttered and honked across the squares and grasses on an historic morning early in the year 1902. It was impossible for any horse-loving officer to protest against this outrageous mechanical intrusion. For, sitting in the car, upright and stern, was the General Officer Commanding, General Sir John French, with the white feathers of his helmet blowing in the wind. Also in the car sat two other officers and, of course, the army's first chauffeur. Sir John French was widely respected for his ability to express his feelings in strong, volcanic language, but on that morning he was silent. However, everyone in The Camp guessed what his thoughts were as a cavalryman renowned for his exploits in the saddle. In the Boer War, as commander of the cavalry division in Natal, French won the Battle of Elandslaagte. Again he led the cavalry in the battles of Reitfontein and Lombard's Kop and at the relief of Kimberley. Obviously he was more suited to a horse than to a motor car.

The introduction of the first motor car into The Camp had been preceded by the following letter to French from Lord Stanley, President of the War Office Mechanical Transport Committee:

'I am directed by the Secretary of State for War to inform you that it is proposed to purchase a motor car for your use in order to facilitate the inspection of works in progress in the district under your command, and also in order that this class of vehicle may be tried as to its suitability for aiding, by its capabilities of rapid locomotion, the command of troops in the field.

'The Secretary of the Mechanical Transport Committee has informed me that you consider that a suitable class of car would be one which would carry six persons, travel over rough and bad road, and surmount very considerable slopes. Such a car would be somewhat heavy, and would of necessity be one of very considerable horse-power.

'It has been observed that, although the tendency is to increase the horse-power of cars, yet the heavy type of car carrying a considerable number of passengers is giving way to the "light car" carrying at most four persons, including the driver, and it is thought that though the heavy car would possibly be the better for visiting works, yet the light car will probably be found the better for the use of a general commanding troops in the field.

'It is therefore proposed to at first experiment with a car of this description, and such a car is being obtained and will be dispatched to you as soon as possible. I am to ask you that you will give it a thorough trial in both the above-mentioned uses, and report as soon as possible as to its capabilities, merits and defects.'

At the time such an experiment in the use of the motor car was regarded as novel and, indeed, revolutionary. However, there were members of the War Office Mechanical Transport Committee who argued that one day the automobile, in variegated forms, would equal if not surpass the horse in military importance and usefulness. They noted how in the Gordon-Bennett Cup Race of 1901 the intrepid driver Monsieur Girardot had sped to victory at thirty-five miles an hour. On the other hand, highly respected officers such as General Sir Evelyn Wood, whom we met in Chapter 6, saw no chance whatsoever of mechanical vehicles ever replacing the horse in warfare. He thought motor cars, bicycles and suchlike were ingenious and interesting, yet obviously inferior to the horse in reliability and manœuvrability.

Not for long was Sir John French's motor car the only one

in The Camp. Soon the twentieth century was in its Aldershot infancy making its foreboding presence known through the cranking of engines, exploding exhausts and the honk-honk of horns. Wealthy young subalterns began to buy motor cars, not for the 'command of troops in the field' but for fun. Goggled, and with white dust coats covering their knickerbocker suits, they drove from The Camp into the town and into neighbouring Farnham, alarming the horse traffic and delighting small boys. They liked to add to the glamour of their pastime by driving with girls at their side; girls with hats made secure from the wind by the covering of a silk or cotton cloth, knotted under the chin. For some time such mad-cap 'joy-riding' by girls with young officers was condemned as 'flighty and unladylike'. Motoring was looked upon as a dirty, unhygienic pastime also, because the cars raised so much dust from the unpaved roads, covering the joy-riders in dirt with shreds of horse-droppings. However, there were pioneer thinkers at the War Office who stressed that officers should make motoring their hobby, for the more they knew about the automobile, the more suited they would be for the mechanisation of army vehicles which was surely on the way.

Within a few years even more revolutionary inventive influences were to affect Aldershot. In December 1903, at Kitty Hawk, North Carolina, the brothers Wilbur and Orville Wright made four flights in a glider fitted with a petrol engine, the longest of these being 852 ft. and lasting fifty-nine seconds. The heavier-than-air craft was now being born. Investigation of these Wrights revealed them not as highly educated engineers or mathematicians, but simple builders and repairers of pedal bicycles. They had received only an elementary education at a free school in their birthplace of Dayton, Ohio. The wonders being accomplished by the Wright brothers caused the progressives of the War Office and Aldershot to forget General French's experiment of commanding troops from a motor car. Admittedly the military value of a heavier-than-air flying machine might be doubtful. However, bold War Office thinkers now believed that the army should welcome and encourage talented inventors, whatever their formal education or social background. In this movement towards mechanisation their might be in Great Britain, as well as in the United States, simple men like the Wrights with brilliant ideas or inventions to contribute to our armed forces. The identity or social class of the genius did not matter. Only his ideas and creations were of importance. This was the philosophy which brought to Aldershot a man who attracted a far greater stir than Sir John French's motor car.

This man, when first seen wandering about The Camp, aroused curiosity and some suspicion. In appearance he was the very antithesis of the British officer and gentleman. Adorning his cheerful but proud face were moustaches which thinned into long waxed antennae, and a goatee beard. But even more unforgivable was his dark hair. It was long and flowing, like the tresses of a girl. And topping the hair was a cowboy hat with a wide brim. Sometimes his breeches suits and stockinged legs would terminate in high glossy boots with silver spurs. He spoke with that American accent known as 'the Texan drawl'. His name was Samuel Franklin Cody, generally known as 'Colonel Cody'. At both Aldershot and Farnborough this strange-looking man was, it was noticed, associating with some officers of distinction. Nevertheless there were other officers of strict conventional rigidity who recognised this Cody as 'an outsider', and denounced him as one. Lower ranks generally regarded his appearance with approval, but of course they thought him a freak.

It was whispered that this Cody was a genius—which might explain his long hair and generally wild appearance. He had, it was said, been sent to help the army in the development and manufacture of man-lifting kites which would be invaluable for aerial surveillance of the enemy behind its own lines. Men in kites would be able to spot enemy concentrations, and guide the artillery into breaking up such concentrations. Indeed, this genius had been lured by the War Office from the Royal Navy to whose command he had been revealing his kite secrets. Then, looking hard at Cody, soldier after soldier identified him. He was none other than 'Colonel' Cody, trick rider and crack marksman, who, with wife and sons, had performed at fairs and in music halls all over the country. No wonder that those officers who had regarded Cody with suspicion and disapproval since his first appearance in The Camp became still more critical. They denounced him not only as 'an outsider', but also as 'a bounder'. They began to treat him with devastating rudeness. When he was introduced to them as 'Colonel Cody', they would reply, 'How-do-you-do, *Mr*. Cody, with loud emphasis on the mister. Cody never winced, but he was pained and upset. Like most Americans, he hated to be disliked. Yet this long-haired and goatee-bearded 'outsider' and 'bounder' was one day to die in Aldershot to the great grief of the army and to be buried in its military cemetery with every branch of the services represented.

The almost incredible story of Samuel Franklin Cody began at Birdville, Texas, where he was born on March 6, 1861. His father was the grandson of an Ulsterman and his mother had sprung from Dutch peasant stock. These had a ranch where they raised a daughter and three sons of whom Samuel was the youngest. The Cody boys were from infancy brought up with affection but, at the same time, with great toughness. They had to be strong and rough and daring, for the Cody ranch was situated in the middle of hostile Red Indian country. Daring members of the Sioux tribe would from time to time attack the ranches of white men with murderous cunning and ferocity. They would in these raids steal cattle and horses, and if they got the chance they would also kill every white man and woman they could find. Thus, from the age of ten, Samuel Cody was a fine horseman and a crack shot. When the redskins were spotted lurking on or near the ranch, father Cody would

mobilise his forces with a cry of 'At 'em!' Then father, three
sons and ranch hands would grab firearms, mount their horses
and ride away in pursuit of the enemy—as in a modern cowboy
film.

Samuel never even saw the manufactured toys which
children played with in the northern states. Nor did he play
their imaginative games. Everything this boy owned was real—
his horse, his rifle and pistol, his ammunition, his lasso. He knew
no Sunday suits or formal haircuts or even formally laid
dining-tables. But in his tough, unconventional childhood
Samuel did adopt a pastime which must be mentioned. For this
pastime influenced his entire life. One day the ranch's Chinese
cook made Samuel a kite with a big ball of string attached. It
was a typical burning hot Texan afternoon, but nevertheless
after the initial wavering the kite rose high in the sky to the
rapt wonderment of the small boy. Samuel had to work hard
and to ride hard on the ranch. But whenever he could steal
some leisure it was usually to design or to fly a kite.

Meanwhile the Cody family's sporadic warfare with the
Sioux Indians continued. To these Indians the father and three
sons, always zestfully ready for battle, were the worst palefaces
in all Texas. Cody was in his teens when one day, suddenly, a
powerful Sioux force took the ranch by surprise. In the thick
smoke and dust of the fight that followed, Samuel became
separated from the rest of the family. The ranch house—the
Cody home—was burned down. A bullet struck Samuel, who
thereupon hid under a mass of debris. When later he emerged
and limped to the smoking ruin of his home he could see
nothing of his parents, brothers or sister. He believed they had
met the usual fate of captives of the Sioux—death by torture.
The wounded boy crawled and swam over land and streams
until he reached the tiny U.S. army encampment at Fort
Worth. When, as the weeks passed, still no news came of his
relatives, Samuel bravely accepted it as a fact that they were all
dead. He became a professional buffalo hunter and after that a

cattle ranger. And several years later he discovered that his parents, brothers and sister were alive after all. They had gone to live in another part of the state after searching vainly for Samuel and then deciding that he had been murdered by the Indians.

After a variety of adventures, which included participation in the Klondike Gold Rush, the young man became a horse dealer. He made trips to England with equine cargoes. There he became friendly with John Blackburne Davis, supplier of horses to King Edward VII. Samuel went to the Derby as guest of this bloodstock expert and was introduced to his daughter Lela. This Lela, in addition to being a girl of intelligence and beauty, was a bold and intrepid horsewoman and also a crack shot. She shared the Texan's passion for fast riding, for trick shooting with rifle or pistol. They married, and their partnership became practical and perfect until, later in their lives, there were difficulties and strife. Soon after the wedding relatives and friends of Lela were regretting that she had married an American who was proving to be 'an adventurer'. Samuel had taken his bride to America, and a dreadful rumour trickled from there back to England that she, a gentlewoman, was participating with him in kind of 'Wild West circuses'.

This was true. Cody was touring America with Forepaugh's Wild West Show. In 1889 his wife was galloping around the arena with him, firing dummy bullets from her gleaming silver pistol in mock battles with Red Indians. The horsy English lady who in England was on friendly terms with King Edward and other members of high society was now 'mucking in' with a rough mob of bronco-busters, entertainers and their noisy females, moving from town to town across the United States in crowded, insanitary caravans. Samuel had a namesake, friend and rival in the famous Wild West entertainer Buffalo Bill, who in private life was William Frederick Cody. Fifteen years older than Samuel and no relation, Buffalo Bill had been his hero since boyhood. Now Samuel was fulfilling youthful

ambitions by emulating Buffalo Bill in show business. When
Buffalo Bill made a triumphal tour of the British Isles Samuel
Cody decided to follow him there with his own show.

He arrived in England as Colonel Cody, self-styled. And
years of success followed in Britain and throughout Europe for
the 'colonel', his lady and their troupe. With them were two
sons, born in America, and later a third son was born on the
Continent. First the Cody Wild West show toured English
music halls in exhibitions of trick riding and shooting. The
show moved to the Continent and it was greeted outside the
Casino de Paris by a vast and hysterical crowd, crying *Vive
Cody!*' They appeared in North Africa, both Cody and his
wife completely outclassing the style and prowess of the much-
vaunted Arab horsemen. Their sons spent hours daily in the
saddle from infancy. And the boys joined Cody in a much-
publicised revival of chariot-racing on a Rome racecourse. In a
contest of six teams of chariots Cody and his son Vivian, then
aged ten, were the winners.

Back in England, the Cody act toured the music halls with
performances of increasing sensationalism. At the Alhambra in
London the boy Vivian accurately fired at targets while his
father swung him upside down by the ankles. Cody placed a
cigarette between the lips of his wife Lela. Drawing his
revolver he shot it from her mouth. Next Lela, shapely in pink
tights, stood in a frame of wooden balls. The husband, firing
his rifle, shattered each ball, one by one. Then the fantastic
long-haired Cody, in huge cowboy hat, scarlet shirt and
buckskin breeches, seized his wife's hand and kissed it while the
audience roared its cheers.

Suddenly Cody decided to become a 'serious' actor-play-
wright. He shut himself up in a room, writing furiously. The
result was the Western melodrama *The Klondyke Nugget,*
followed by *Calamity Jane* and *Nevada.* Today these works
would be ridiculed by both public and critics as 'corny'.
Samuel Cody usually played the role of the sinister adventurer

and Lela the pure and innocent maiden whom he planned to enslave and deflower. Twirling his moustaches, Cody strutted the stage with villainous invective and hisses, while Lela, in a tremulous, high-pitched English voice, won the audience's indignant sympathy. Shooting and even live horses were features of the melodramas. In 1898 Cody brought his blood-curdling *The Klondyke Nugget* to the Theatre Royal, Aldershot. Large numbers of soldiers and their wives crowded into these performances, little realising that the villain of the play would one day die as the beloved hero of The Camp.

Small wonder that sophisticated English people regarded the so-called Colonel Cody as a preposterous character, despite his skill as a horseman and his brilliance as a shot. The army, stirred by stories of Cody's shooting prowess, sent representatives to exhibitions of riding and marksmanship which he gave, from time to time, in the open air. These observers watched with amazement as the mounted Cody hit target after target with rifle or pistol at full gallop. They agreed that the army had no marksman to equal him. They suggested that Cody might be appointed a civilian instructor to the troops. But the War Office, prejudiced against Cody as a flamboyantly attired American and hissing stage villain, decided that such an appointment would damage the army's dignity, so refused to consider employing him. This was before the adoption of a freer official policy towards outsiders and eccentrics who might be useful to the forces.

While touring the British Isles with his melodramas, Cody started to revive his boyhood pastime of making and flying kites. But now the kites were enormous—and designed to carry live men. Cody, Lela and two of their sons each made ascents at Blackpool to the amazement of hundreds of holiday-makers on the beach. One day soon afterwards in North London Cody rose 80 ft. in a kite, only to be blown into a tree by a sudden gust of wind. He fell from the kite, grabbed a branch of the tree and swung like a monkey until rescued. The stage villain's dramatic experiments with man-lifting kites on the coast were naturally observed by members of the Royal Navy, and finally the Admiralty decided that he could be ignored no longer. So the Navy invited Cody to demonstrate his kites under its auspices at Portsmouth. The entire Cody family participated, and Leon, the eldest son, rose more than 800 ft. in a kite. Later Cody was making kite flights from warships, snapping photos as he was billowed about in the sky, to prove the usefulness of his craft in aerial reconnaissance.

It was Cody's success with the Royal Navy that caused the army to invite him to become attached to the British War Department's balloon factory and school at Farnborough. There he was encouraged to continue with his kite experiments and also to pour out his ideas on the future of military balloons. He formed a close friendship with the new commandant of the balloon factory, Colonel John Capper. Capper, who in a brilliant career was to rise to a major-generalship and to be knighted, was enchanted by this self-styled colonel with the long flowing hair and cowboy outfits. Until they came on Christian-name terms, Colonel Capper always respectfully addressed him as 'Colonel Cody'. However, stuffier members of the army were incensed by the flamboyant stage villain and his kites. Cody would, to the delight of privates and N.C.O.s, appear on Laffan's Plain astride a white horse which would go into a full gallop while he fired bullets in the air from a silver revolver. And he would invariably wear oversize silver spurs, reputed to be of great value. Despite Capper's enthusiastic patronage of the Texan, other officers treated him with cold disdain. Capper tried tactfully to persuade Cody to be a little more conservative and restrained. Since so many officers were displeased by Cody's long hair, Capper got him to knot it into a bun at the top of his head and thus be concealed under his spacious cowboy hat.

The sheer goodness and honesty of Samuel Cody gradually conquered the prejudices of convention and class. Aiding Cody

in this conquest was his genius as an inventor and the fearless-
ness of him and his sons in their many kite ascents over Farnborough and Aldershot. Indeed, many who had insisted he
was a mere mister now cordially addressed him as 'colonel'. On
the formation of the Kite Section of the Royal Engineers, Cody
and two of his sons were appointed as instructors. In 1904 and
1905 Cody was a picturesque figure among officers in army
manœuvres. In 1906 he was officially designated Chief Kiting
Officer. Soon he was being showered with congratulations for
fitting a 15 h.p. engine to a kite which, without a pilot, sped
through the air for more than four minutes. Experts predicted
with confidence that within two or three years men would be
flying long distances in engine-powered kites. They said that
the world was moving into the 'kite age'.

The 'kite age' never came after all. Colonel Capper of the
Farnborough Balloon factory asked Cody to divert his attention towards the completion of a dirigible balloon. This, a
product of the factory, was Nulli Secundus I, which had a
length of 122 ft. and a diameter of 26 ft. Cody designed the
craft's understructure and fitted a 50 ft. French engine, with
propellers designed by himself. After a series of mishaps it
seemed that Nulli Secundus I was a success. It cruised over
Aldershot and district while crowds below craned their necks
and waved handkerchiefs. Then one day in 1907 Capper and
Cody took her over London. To the wonder of King Edward
and Queen Alexandra Nulli Secundus I glided over Buckingham Palace and its garden; and next over St. Paul's and the
City at sixteen miles an hour. But when Nulli Secundus I tried
to return to Farnborough a high wind drove her away. When
she tried to make a landing on Clapham Common the gathering of a vast crowd caused Cody and Capper to try elsewhere.
Eventually an emergency descent was made in the grounds of
the Crystal Palace. There, bedded down with sandbags, this
flying wonder remained for several days until heavy rains and
gusty winds damaged both her envelope and understructure.

Thereupon she underwent an humiliating deflation from the knife of a Royal Engineers sergeant. She was, with gross indignity, transported back to Farnborough in army transport wagons.

Meanwhile Cody became increasingly impressed by the continued flying feats and experiments of the Wright brothers in his native America. With the support of Capper, Cody begged the War Office to allow him to make heavier-than-air machines. The War Office had rejected an offer for the purchase of the Wrights' plane, and, indeed, had become deeply sceptical about this new-fangled invention. Nevertheless, in a burst of generosity, the War Office made Cody a grant of £50 and told him to go ahead and make his flying machine. In 1907 Cody began his task. He could not afford a new engine, so he borrowed the one from Nulli Secundus II, an airship which was to have replaced the unhappy Nulli Secundus I. Hours daily he was busily putting his plane together in the privacy of an Aldershot shed while press and public remained sceptical that he could succeed. It was pointed out that Cody, far from being a trained engineer, was merely a trick horseman and marksman, and an actor of melodrama. He had been lucky in flying kites, but surely he was not the man to build and fly a machine which would be too heavy to float in the skies? There followed for Samuel Cody a period of humiliation when his popularity, won after so many snubs, was quickly evaporating.

He wheeled his flying-machine—named British Army Aeroplane No. 1—from the shed one day in 1908. It looked so fragile that watchers wondered how it could possibly carry its inventor, who weighed 18 stone. Cody took his place at the controls. Space was cleared of people for the pilot to speed the plane into a run and to an ascent. However, British Army Aeroplane No. 1 refused to rise from the grass. Rain was falling hard. As Cody, dripping and disconsolate, tinkered in vain with a dead engine, a jeering crowd converged upon him and his masterpiece. There was uproarious laughter and cries of

'Try a motor bus!' The plane was pushed back into the shed. Newspaper comments next day were scathing. But on May 16, 1908, Cody really flew. A tablet on a tree at Farnborough commemorates the feat in these words:

'Colonel S. F. Cody picketed his aeroplane to this tree and from near the spot on 16th May 1908 made the first successful officially recorded flight in Great Britain.'

On this flight Cody had travelled 50 ft. with the plane's wheels above the ground. Happy and elated, he made four more flights that day, one over a distance of 150 ft. at a height of 10 ft. But his fourth flight ended in his plane striking a water trough, smashing many of its delicate spars and ribs.

At last Cody had really flown a flying-machine. But the British would not take the American seriously. The many who made a big joke of 'Colonel Cody' were egged on by comments of newspapers who labelled Cody 'the bluffer', 'Yankee showman', and his plane 'the mowing machine' and 'the grasshopper'. Trying to ignore the nation-wide insults and jeers, Cody went doggedly on with his aeroplane flying. One day when he had mastered the art of turning in the air, the Prince of Wales (later King George V) drove over from the Royal Pavilion, where he was staying, and watched. Soon the prince was vigorously shaking the American's hand and repeating in his hearty voice, 'Congratulations' and 'Well done'. They had a long conversation, with the prince addressing Cody as 'colonel'. The prince had, indeed, more foresight regarding aviation than the War Office, and he still retained his friendship and affection for Cody when he became king.

As time passed, Cody's aeroplane experiments cost the government more than that initial grant of £50. Indeed, the information leaked that his experiments, together with those of another distinguished flying pioneer Captain J. W. Dunne, had cost the nation £2,500 in one year. Critics asked how long such extravagant folly would be allowed to continue. The Minister for War of the time was Richard Burdon Haldane (later Lord Haldane) who is remembered today for his many admirable army reforms. Haldane travelled down to Aldershot to see Cody. Gently Haldane told him that his services were no longer needed because the War Office, after much careful consideration, had decided that aeroplanes were quite unsuited for military use. The aircraft for war, emphasised the wise Haldane, was the airship—not the flying machine. But the Minister for War made a generous concession to Cody. This was that he would be allowed to use Laffan's Plain and the Aldershot and Farnborough areas for any flying experiments. Moreover, Haldane added that Cody would be allowed to keep the flying-machine he had constructed—but with one reservation. The engine of this craft must be returned to the War Office because it was needed for re-installation in airship Nulli Secundus II.

A newspaper, believing that this was truly the end of the comic 'Colonel' Cody, snapped a 'farewell' photograph of him leaving the Farnborough Balloon Factory, printing it with the caption: 'Cody's Farewell. God save the Taxpayer. We could not love you, sir, so much, loved we not money more! The War Office has parted with Mr. Cody, and we publish above a photograph of his pathetic if somewhat undignified exit. It will be hoped that the Authorities will now finance an *Englishman's* Experiments.'

The newspaper, like Haldane, lacked foresight. Its editor did not realise what would be the reaction of the rank-and-file of the army to the War Office's sacking of Cody, or that the British public would soon hail him as its hero. Now Samuel Cody was truly an outsider as far as the War Office was concerned. But practically nobody in Aldershot regarded him as a bounder any more. Progressive-minded soldiers retained their admiration for him and were pleased that he would at least still be among them. They were glad to let him build a personal shed and workshop on the edge of Laffan's Plain. They enjoyed his flamboyance, the flamboyance that had once

M

caused such annoyance among stuffy members of the Forces. On his part, Cody made generous concessions to local conventionality and conservatism. He had his hair cut. He now wore a British fedora more often than the Texan cowboy hat. He also abandoned the rest of his cowboy attire for well-tailored English suits.

He bought an engine to replace the one that the War Office had taken away and he began to fly again. Soon he was making longer flights, higher and higher from the ground. He took up Colonel Capper as the first passenger ever to be carried in a flying machine in Great Britain. Later another passenger was General Sir Horace Smith-Dorrien, then the Aldershot G.O.C. One early September morning Cody soared over Aldershot, Fleet and Farnham for 65 minutes, attaining a record height of 800 ft. Square-bashing soldiers on early morning parades broke ranks to mill around, cheering and waving weapons and headgear. Civilians of Aldershot and neighbouring towns soon realised that Cody was breaking records. Crowds rushed pell-mell to Laffan's Plain to mob him when he landed. Yet Cody, so often the exhibitionist, was so moved by this welcome that he ran into a shed and hid. The flight won him the silver medal of the Royal Aeronautical Society.

At last the British had really taken this American close to their hearts. He was so grateful and so touched that he wanted to make every possible gesture in reciprocation. The supreme gesture came when he was in Doncaster to compete with his plane in an aero rally in October 1909. He publicly renounced his American citizenship and became a British subject. The naturalisation ceremony was organised and advertised by Cody as if he were in one of his elaborate show productions. It took place outside the huge shed in which his plane—nicknamed 'The Flying Cathedral'—was resting. Over this shed was a flagstaff from which was displayed a large U.S. flag. The Doncaster town band, hired for the occasion, played the

American national anthem while Cody stood at attention. When the 'Star Spangled Banner' ended Cody, with the town clerk at his side, made his declaration: 'I, Samuel Franklin Cody, do swear that I will be faithful, and bear true allegiance to his majesty the king, and his heirs and successors according to law.' With a flourish he signed the naturalisation papers. The American flag was lowered. The band played 'God Save the King', with Cody at the salute, while the Union Jack was raised on the flagstaff.

Thereafter Samuel Cody was no longer looked upon as the alien freak. He, with his series of planes and many achievements in the now fast-expanding aviation world, became a familiar part of the English scene. Even the stingy War Office, upon legal pressure, awarded him £5,000 in recognition of his earlier pioneer work with the man-carrying kites. In 1910 and 1911 he won the premier British flying award, the Michelin Trophy, for flights of 195 and 242 miles. In 1912 Cody was handed the prize of £5,000 for winning the great army aeroplane tests on Salisbury Plain in competition with the world's best airmen and machines. For by then the War Office had, in a change of mind, decided that the flying-machine had some future in warfare.

In the summer of the following year Cody, now aged fifty-two, was busy preparing for entry in the Northcliffe competition for a seaplane flight around Great Britain for a prize of £5,000. He had designed and built a new biplane which, he boasted, 'flies like a bird'. He was daily flying this plane around Aldershot before taking it to Calshot for floats to be fitted for the race. On the morning of Thursday, August 7, 1913, he went for a flight in his plane with an officer friend of the 20th Hussars. 'You see,' Cody boasted to him, 'I can't fail to win in this. It flies to perfection.'

Later that morning Cody met his friend W. H. B. Evans, of the Indian Civil Service. This Evans had been the Oxford University and Hampshire cricket captain. 'I'll take you on the perfect flight in the perfect plane,' Cody told Evans with a laugh. Cody's two elder sons and others watched his plane ascend from Laffan's Plain and fly off towards Bramshot. Then it was seen returning to land again. While about a mile away from Laffan's Plain, the machine seemed to be doubling in the centre. Wings broke away and fell into the trees; and seen falling too were Cody and Evans. The bodies were found on the ground only two feet apart. Cody was in the white dust coat he always wore while flying. Friends of Cody gave a start when they saw the socks of the dead Evans. They were bright green. The friends remembered that Cody had a horror of green as a colour certain to bring him bad luck. He never wore anything green while flying. He refused to take up anyone in a plane if there was green in his or her clothing. Apparently he had not noticed the colour of his friend's socks.

But Evans's green socks were not so much the cause of the disaster as was the engine of the biplane that 'flies like a bird'. Experts, examining the wreckage, suggested that the engine had been too heavy for the fuselage. The plane had fallen from a height of between 300 and 400 ft., but the two men had dropped from it at a lower altitude than that. They might have lived, said a witness at the inquest, if they had been strapped in the craft. Cody had died from a broken neck. At the inquest people were sobbing. Indeed, the death of Cody threw into grief nearly everybody who had known him or even seen him. For throughout The Camp, Aldershot and Farnborough, Cody, the original, and the eccentric and nonconformer, had become greatly beloved. Even today elderly men of the district who knew or merely saw Samuel Cody in their boyhood recall with emotion those days of his death and burial.

The man who on his arrival in Aldershot had been denounced as 'outsider' and 'bounder' was given the biggest funeral in the history of The Camp. The Aldershot G.O.C. was now Lieutenant-General Sir Douglas Haig (later Field Marshal Earl Haig), a stickler for convention, an advocate of

always doing 'the right thing'. Friends of Cody were rather
diffident in approaching him with the request that this civilian
with the spurious colonelcy should be granted the posthumous
honour of a military funeral. They recalled that soon after
arriving in Aldershot Haig sent his aide-de-camp to Cody
with a message that he wanted to see him. Cody, in honest
ignorance, had replied: 'But who is this General Haig? I've
never heard of him—but no doubt he has heard much of me.'
Nevertheless Haig was impressed by Cody when they met.
Haig was also aware that his friend King George V had always
addressed Cody with the prefix of 'colonel', which surely made
it a legal service commission. So Haig granted the request for
a military funeral for Cody, but quite properly issued a
Command order that there was no compulsion for troops to
attend the ceremonies.

Nevertheless, nearly all the troops of The Camp turned up
at a procession and funeral watched by 50,000 civilians from
London and from many miles around the Aldershot district.
The coffin was borne on a gun-carriage supplied by the Royal
Artillery. On the coffin was a huge wreath from the widow, its
flowers shaped into the steering wheel of an aeroplane with a
broken shaft. Cody's three sons, Leon, Vivian and Frank,
walked behind the gun-carriage. Then came a variety of
detachments and servicemen, marching in slow step to the
tunes of the pipers and band of the Black Watch. There were
300 members of the young Royal Flying Corps, and naval
flyers. There were representatives of every regiment, corps and
department of the Aldershot Command. The procession
started at the Cody home at Ash Vale and moved through
streets where all shops were closed and shuttered and where the
blinds of all houses were down, to the military cemetery at
Thorn Hill, Aldershot. At the graveside committal the Black
Watch pipers played 'Lochaber No More'. The faces of the
troops on parade were expressionless, immovable. But the
hundreds of men, women and children who stood watching

and listening in the cemetery wept without restraint. Later a sculptured figure of Christ was raised over the tombstone.

Thus was the climax of the life of the small boy who was taught to fly kites by a Chinese cook on a ranch near Birdville, Texas, U.S.A.

9

King George's Friend

Probably the happiest period in the history of The Camp and Aldershot town was the five years from 1907 until 1912 when the G.O.C. was Lieutenant-General Sir Horace Lockwood Smith-Dorrien. Despite the conventional expression of regrets of traditional military farewells, no one was really sorry when in 1907 grumpy, peppery Sir John Denton Pinkstone French left the Aldershot Command to become Inspector General of the forces. His successor Smith-Dorrien, then aged forty-nine, seemed to everyone to be a less difficult and complicated character. Here, in comparison with the complex French, was a happy extrovert, effervescing with the joy of life and with goodwill not only towards all men but also towards all horses. But there was nothing overtly sentimental about the vigorous-looking, square-jawed Smith-Dorrien. His eyes and bearing with his strong, decisive manner of speaking bore witness of self-confidence and resolution.

Smith-Dorrien, in the language of 1907, was a gentleman, sired by the right father (Colonel Robert Algernon Smith-Dorrien, J.P., of Haresfoot, Herts) and educated at the right places (Harrow and Sandhurst). Substantial 'private means' doubtless contributed towards the generosity of Smith-Dorrien's temperament. In contrast, predecessor French was periodically tortured by financial worries. Indeed, French had once in India obtained a loan of £2,000 from brother officer Douglas Haig who was to succeed him as commander of the British Army on the Western Front in the First World War. And he was not in a position to repay it for a long time.

When Smith-Dorrien left Sandhurst in 1877 he was gazetted a lieutenant in the 95th Foot. Two years later he was in the thick of the Zulu War. Next he was in Lucknow, India, to distinguish himself not so much in soldiering but as a crack contestant on racecourse and polo field. He had evolved an absorbing passion for horses, not merely as steeds but as individual characters and personalities. Back in England at Staff College Smith-Dorrien became a popular Master of its Draghounds. Then he was in India again with his regiment and with a string of beloved polo and racing ponies. Kitchener called him from these pleasures to Egypt where he commanded a Sudanese battalion in the Battle of Omdurman. Next he was fighting in the South African War until Field-Marshal Lord Roberts sent him to India in 1901 as Adjutant General.

Smith-Dorrien was touched by the fact that while British officers in India had a lot of fun, the fate of the ordinary soldier was usually dull and dreary. Thus through his enthusiastic initiative clubs and recreation grounds were established for the troops in every station in the country. In 1902 Kitchener became commander-in-chief in India and appointed Smith-Dorrien to the command of the 4th Division in Quetta. The 4th Division quickly won a reputation for happiness—and efficiency. Both officers and men adored their commander and followed with enthusiasm his exploits on his horses. The entire division celebrated when on one single day at a race meeting Smith-Dorrien won two steeplechases and one hurdle race. It was said that this 4th Division commander not only knew the faces, the characters and idiosyncrasies of every one of his officers, but also those of their ponies. Then Smith-Dorrien was interested in ordinary soldiers as personalities and not merely as the faceless automata of 'other ranks'.

No wonder that when John French passed on the Aldershot Command to Horace Smith-Dorrien the atmosphere among its troops became less tense and increasingly relaxed. And under the influence of Smith-Dorrien the conduct of non-coms and privates in the streets of Aldershot improved vastly. Until then the bars, cafés and music halls were rowdy places, particularly on the evenings of military pay days. The town was patrolled by some 500 picquets (military pickets), who, far from preserving order, often contributed substantially towards disorder. For example, there would be an altercation in a public house. Someone would rush out into the street and report this to the picquets. The picquets would thereupon plunge into the public house and attempt to seize the soldiers causing the disturbance. Such soldiers together with civilian allies would, in their inebriated condition, unite in resistance to the picquets until the public house was a shambles. Stories of such battles, in bars and open streets, appear in abundance in the local newspapers of those unruly times.

In studying details of recurring Aldershot disorders Smith-Dorrien decided that the well-intentioned picquets were provocations rather than keepers of the peace. He resolved that all should be withdrawn and that an appeal would be made to the honour of his troops instead. Officers with old-fashioned ideas were horrified that the town of Aldershot should be left without the protection of picquets, always watchful and ready to hustle offenders to the guardroom. They predicted chaos for the town, the destruction of property, assaults on innocent citizens and the violation of pure women by the undisciplined soldiery. However, the commander's order withdrawing the picquets and placing his men on their honour to behave with reasonable propriety had an amazing effect. Certainly excessive drink and squabbles over women still caused incidents in the bars, but without the intervention of picquets these would more easily and frequently fizzle out. The troops appreciated the trust now placed in them by Smith-Dorrien and tried their best to live up to this.

Smith-Dorrien noticed that the army was attracting many more recruits of refinement than in the past. A substantial proportion of the soldiers aspired to more diverse pleasures than booze and women. Thus he enthusiastically encouraged institutes and clubs where those pleasures were not dominant. He was a great admirer and friend of the Daniell Soldiers' Home and Institute. Louisa Daniell and her daughter were dead, but their work at the institute was being carried on by a splendid honorary superintendent named Kate Hanson. Smith-Dorrien was pleased when the new Wesleyan army recreation centre opposite Corunna Barracks, of which he laid the foundation stone, was officially named the Smith-Dorrien Methodist Soldiers' Home.

On May 6, 1910, King Edward died, to be succeeded by George V. At that time Aldershot was becoming a quieter, more restrained place, as if turning towards the finer virtues as exemplified in the characters and conduct of the new king and queen. Tougher police methods, together with the healthier interests of many recruits, led to a noticeable decline in the number of prostitutes operating in the town. The dance halls with bedroom accommodation had vanished. There were still a large number of public houses, but these did not provide customers with the 'other arrangements' which so scandalised Louisa Daniell and her daughter. However, there still remained outside the town limits near Ash Vale a community of prosti-tutes in shanties and tents. There also was an underworld of unlicensed vendors of cheap liquor, thieves, gamblers and the more disreputable types of gypsies. The police generously left these wild and lawless people alone so long as they kept out of the centre of Aldershot. For a belief still persisted that there could never be an army camp without its horde of low 'followers'.

People are still living who recall the happy, wholesome social life among soldiers and respectable local girls in the early years of the twentieth century in Aldershot. These were the years of the Sunday pastime known as 'walking out', or as 'courting' if the male had matrimonial intentions. In 'walking out' the girls wore outfits strictly reserved for the Seventh

Day and rare social occasions of supreme importance. Some actually faintly powdered their noses, although the use of powder was criticised as 'flighty'. The use of rouge by maidens of the working classes and middle classes in Aldershot was rare. Rouge—and mascara too—was associated with 'fast' London actresses, divorced women and other questionable types. Among respectable girls there was a fervent belief in the qualities of soap as a beautifier. A popular and widely advertised soap in Aldershot was 'Goodwin's toilet soap, a friend in need to all who would maintain their healthy naturalness and beauty of complexion'. It cost twopence a tablet. While eschewing the powder and paint of fast and flighty women, respectable girls would, however, add to their sweetness and glamour by concealing on their persons scented sachets and by surreptitiously popping into their mouths alluringly perfumed cachous. But alas, old folk who remember the Aldershot of those days say that the appearance of many young women was spoiled by decaying or missing teeth. Dentists who had no training or qualifications would too often treat a case of toothache by dragging out the bothersome molar instead of drilling and filling its cavity.

Certainly many cases of 'walking out' between soldiers and local girls led to serious courtship and eventually marriage. Nevertheless early in the century there seemed to be a great excess of spinsters in Aldershot, despite the presence of thousands of bachelor soldiers in The Camp. Excitement was caused among the spinsters in 1911 by a widely publicised story from Canada that '5,000 British wives' were wanted 'for lonely bachelors' in that dominion. This was followed by an advertisement from a firm of Aldershot shipping agents stating that 'we shall be glad to advise and help any young lady who would like to go to Canada'. The offices of this agent became swamped by young women impatiently demanding particulars. A local nonconformist minister made this scene the subject of a sermon. He accused local young men, military and civilian, of wasting their money on the consumption of alcohol instead of doing their duty by marrying and establishing good Christian families and homes. On the other hand elder folk said young men avoided matrimony because of the high cost of living. They referred, for example, to the prices of female clothing in Aldershot shops. A Sunday dress of taffeta cost 17s. 6d., a simple voile robe 7s. 6d. A muslin blouse, a pair of corsets and a pair of voluminous Directoire knickers would each bear the price tag of one shilling.

The scope of the social life of Aldershot soldiers and civilians was also widened in the early years of the reign of George V by the development of bus services. In 1912 the Aldershot and District Traction Company was founded, with the establishment of bus routes to Farnham, Fleet, Farnborough and other towns and villages. Those who had previously 'walked out' in the streets of the town now went on jaunts to places which previously had been almost inaccessible to them. Almost simultaneously arrived the great attraction then known as the Kinema. The first was opened in Aldershot in 1912 and then another in 1913. In 1913 also another variety theatre, the Hippodrome, was opened. These innovations did not affect the live drama of the legitimate theatre which continued to flourish, thanks to tremendous patronage from The Camp. Of course, the troops wanted plenty of sex in the theatre—but they liked also to see virtue triumphant. Typical of such dramatic fare was the play *Two Lancashire Lassies in London* which in 1911 was played to packed and palpitating houses at Aldershot's Theatre Royal. It was about two sweet and innocent girls who came from the virtuous north to win fortune in wicked London. Each was lured by a villain to a life of exciting sin. However, by the last act they had been snatched from the downward path, from 'a fate worse than death', to a virtuous routine of pure, if dull, romance. Another play which filled the same theatre soon afterwards was *The Scarlet Sin* by George R. Sims and Arthur Shirley.

In many respects the lot of Aldershot's 'other ranks' improved under the command of Major-General Smith-Dorrien. Their barracks and huts were made more comfortable. They were being provided with decent crockery. Lights in the recreation rooms and dormitories were strong enough to read under. Of course the better conditions would not have been possible without the support given to the G.O.C. by the Liberal government's Minister for War, R. B. Haldane, who in 1911 became Viscount Haldane of Cloan. For some time, as we saw in the last chapter, Haldane was blind to the military possibilities of the aeroplane. Nevertheless this philosopher-statesman inspired the biggest changes in the army since those of Lord Cardwell in the 1870s. It was Haldane who, against fierce opposition, succeeded in replacing the old militia and volunteer force by the Territorial Army. And it was Haldane who, sensing that a European war was on the way, had worked out detailed plans for a British expeditionary force; also a scheme for an Imperial General Staff.

The Director of Military Training at the War Office between 1906 and 1909, when these ideas evolved into practical planning, was a cool-headed hard-working major-general named Douglas Haig. Both Haldane and Haig were natives of Edinburgh. Both shared a surging desire for changes and reorganisation in the military establishment. Watching him work, Haldane endorsed the opinions of King Edward VII and George Prince of Wales that this officer had great gifts and had a boundless future. In 1912 Douglas Haig succeeded the popular Smith-Dorrien as General Officer Commanding at Aldershot. But first let us review the life of this strange and controversial army commander up to the time of his arrival at The Camp.

Douglas Haig was born in 1861, the son of John Haig of the whisky family, and a descendant of the seventeenth Laird of Bemersyde, Berwickshire. He was difficult and unhappy in his early boyhood, but became pacified and moulded into the conventions of English gentility during his four years at Clifton College, Bristol, from the age of fourteen. There he was looked upon as wholesome but ordinary, with no gifts of scholarship but with a liking for games. Strange that the proud statue of this personage, who attracted no special attention during his schooldays, should now face the Clifton playing fields. When Douglas was eighteen his pious mother died, followed a few years later by his father John Haig. In 1880 Douglas, devoid of any special ambition except to excel at sports, was enrolled at Brasenose College, Oxford. He was pleased to be welcomed by the Master of Brasenose with this inspiring counsel: 'Ride, sir, ride! I like to see the gentlemen of Brasenose in top boots.' The undergraduate obediently threw himself into polo and he played for Oxford in the 1882 university match. He also hunted. He could easily afford to be a fashionable undergraduate because the whisky wealth of the Haigs provided him with a comfortable independent income. While excelling in sports, Haig failed to take a degree. In later life he explained that 'bouts of illness' prevented him from fulfilling the university's residential qualifications necessary for graduation. Yet fortunately the 'bouts' rarely kept him off the hunting and polo fields.

In 1883 Haig became a Sandhurst cadet. He had decided upon the army as a career because of the manifold opportunities it offered for gentlemanly sport—and adventure. At Oxford he had been intent only upon excelling at games. At Sandhurst he resolved to excel at everything. He applied his concentration with equal intensity on books, drill, on polo; but he failed to shine socially or to win many friends. Dashingly handsome, superbly tailored and booted, Haig was one of those men who at dinner parties and receptions seem nervously apart, with no light conversation, jokes or banter. Despite his dash on the sports field, his striking if cold blue eyes and generally manly good looks, he was no ladies' man. He ended his period at Sandhurst first in order of merit, and with the prediction of an instructor that 'before Haig has finished he will be top of the army'.

After Sandhurst Haig served with the 7th Hussars in India. Eleven years later he was in England at Staff College. He had a struggle to get there, despite his reputation for intelligence and his iron ambition. First he failed in arithmetic in the entrance examination. Next it was discovered that he was colour blind. Nevertheless he was given a nomination to the college through the intervention of influential friends. Once inside, he impressed his instructors, played good polo and hunted with zeal. But notwithstanding, the reactions of the fellow officer students to Haig were generally negative. His manners were too aloof, he never tried to be popular, he seemed devoid of the emotions and passions of ordinary men. He did not laugh easily at jokes. Only rarely did he try to make a joke himself. Once at Staff College he was an after-dinner speaker. Mental exercise, he said, could be aided by the sports of hunting and polo. He stressed how topography studied in a practical manner from the saddle of a horse could be more effective than from the mere examination of maps. No wonder that Sir Evelyn Wood became such an admirer of Douglas Haig.

The year 1898 found Haig with the Egyptian cavalry as chief of staff to its commander Colonel Broadwood. He participated in the victory of Omdurman. In between the Egyptian and South African wars he made his first appearance at Aldershot as a brigade major of cavalry. This was when French was in command at The Camp. In 1899 he was off to South Africa. General French was now in command of the cavalry in Natal and Haig was his chief of staff. He was with French in the Battle of Elandslaagte. He returned from South Africa as commanding officer of the 17th Lancers who played brilliant polo. As he raced towards success, this earnest ambitious officer seemed to remain nowhere for long. While in India in 1903 as Inspector General of Cavalry, he moved from his colonelcy to become a major-general a year later.

Vigorously helping to push Douglas Haig to the top was first his loving sister Henrietta—and then the royal family. The wealth and influence from whisky galore made also a contribution to Haig's success. Henrietta was married to William Jameson, whose riches, like that of the Haigs, were derived from a popular brand of whisky. This Jameson, familiarly known as 'Willie', was a member of the fashionable and gilded set of Edward VII who was always attracted by millionaires. Through the Jamesons Haig met and impressed the king. He became an aide-de-camp to Edward and an assiduous writer of confidential letters to Buckingham Palace. From Haig in India, for example, came long letters to the King-Emperor regarding his dusky cavalry and other army matters. Haig was, of course, introduced to George Prince of Wales, who, on becoming monarch, remained a faithful, steadfast friend.

Haig was in the Ascot week of 1905 staying at Windsor Castle while on leave from India. He was introduced to Miss Dorothy Vivian, a maid of honour to Queen Alexandra. A pretty buxom girl, Dorothy had a love for sport, with special admiration for men who excelled at games. She was one of the twin daughters of Lord Vivian. One day Douglas and Dorothy went out on the golf course at Windsor. During a third successive game he waved away the caddies and promptly invited Dorothy to become his wife. They were married less than a month later in the royal chapel of Buckingham Palace. And when a friend commented upon the speed at which Haig wooed, won and wedded, he snapped back: 'And why not? I have often made up my mind on more important problems than that of my own marriage in less time in India.'

Now we reach the stage, between 1906 and 1909, mentioned earlier in this chapter, when Haig worked so successfully and happily with Haldane as Director of Military Training at the War Office. Next he was back in India again, but then as her commander-in-chief, whipping her forces into the state of efficiency and readiness which prepared them for such wide-

spread and successful participation in World War I. It was late
one night in 1911 when a messenger hurried to the commander-
in-chief's quarters with a cable received a few minutes before
and immediately decoded. Haig glanced at the cable, stared
sharply at the messenger and remarked: 'That could have
waited until morning!' The cable was from Lord Haldane. It
read: 'Would it be agreeable to you to be appointed to the
Aldershot Command?'

Smith-Dorrien moved away to take over the Southern
Command at Salisbury. Everybody, both in The Camp and
Aldershot town, seemed sorry to see him go. There was widespread trepidation concerning his successor. Only a few
cavalrymen in The Camp knew Haig, and even they were not
enthusiastic about serving under him. Then Haig caused
offence by insisting upon bringing his own staff with him from
India. Members of this sun-tanned, somewhat supercilious
coterie were for a time regarded as if they were men from Mars,
entirely alien to the scruffy English character of Aldershot.
Resentful officers, and privates too, whisperingly referred to
Haig's staff as 'the Hindoos'. 'What are the Hindoos up to
today?' they would ask one another on seeing Haig ride from
Government House with his clique on horseback behind him.
When anything went wrong in The Camp blame was in-
variably placed upon 'those bloody Hindoos'.

But Haig and 'the bloody Hindoos' meant well. One of
them, as Brigadier-General John Charteris, was after Haig's
death to describe in detail the pleasure which Aldershot brought
to Haig. He delighted at being back in English surroundings,
and in an English home that brought great happiness to his
wife Dorothy. She made the interior of Government House
conventionally upper class, with signed portraits of royalty in
silver frames on mantelpiece and grand piano, with drawing-
room overpopulated by nicknacks of china and silver, with
paintings of horses, dogs and other fixtures of the countryside
on landings and in the study, with riding crops, shabby

N

mackintoshes and a smell of leather in the hall. Dorothy's
pursuits were also those of conventionally charitable gentle-
women. She visited regularly The Camp's hospitals, walking
from bed to bed with crisp words of cheer. She was interested
in the wives and children of The Camp, joining with her
social equals to knit and sew them warm sensible garments and
generally to improve their welfare. She was a charming dinner
hostess, with a gift of putting guests at ease in contrast to the
sometimes embarrassingly stilted conversation or elongated
silences of her distinguished husband. She was adept at making
just the right remarks when frequently, as the general's lady,
she handed out silver cups, shields and other geegaws at army
sports events. The relationship between Dorothy and her
husband was always happy and smooth.

Haig's average Aldershot day started on horseback when,
early in the morning he observed the training of his infantry
and cavalry. He would watch the scene with concentration,
sharply drawing attention to any bungling or inefficiency. If
necessary, he would order an exercise to be repeated again and
again until it was near perfection. Young officers would be
called before him to be lectured or berated. Rarely at this
period did he openly praise smartness or good performance of
the troops. For these, he said, should be normal in the British
army and therefore not worth a comment. He could rebuke,
frighten and mortify without uttering a word. When, at the
unexpected appearance of the G.O.C., a soldier smoking on
duty dropped his cigarette, Haig sat silently on his horse staring
fixedly at the cigarette on the ground as it burned itself out.
Then, without a glance at the mortified soldier, Haig rode
away.

At the end of the early morning drill and manœuvres, Haig
spurred his horse to a gallop across the countryside, usually
with members of his staff—'the Hindoos'—whom he loved to
race or evade, just for the fun of it. From midday until luncheon
Haig would be at his desk at army headquarters immersed in

paper work and in interviews. After lunch he was golfing or
playing tennis until it was time for tea. He was always a self-
conscious devotee of physical fitness and would, from time to
time, go on faddist diets. He would often consume so-called
'health foods'—patented brown breads, iron, nut and milk
preparations. In the early evening Haig read or scanned military
magazines of England, Germany and France. He had little
taste for good literature, for music or art.

Twice a week he and his wife held unpretentious dinner
parties in their home at which he got to know and observed
the more junior officers of his Command. On their arrival Haig
would lurk unseen on the landing at the top of Government
House's white pillared staircase and peep down upon them to
study their deportment. The dinners, at a narrow table in the
long dining-room, generally became natural and relaxed, with
more conversation about sport than of military matters. The
understanding hostess was specially clever at making shy young
men feel at ease. Only occasionally at these 'family parties'
around his own table did Douglas Haig cause strain by relaps-
ing into long silence or, worse still, by becoming inarticulate.

While Haig could write with clarity and precision, he was
often poor at expressing himself in speech. Those close to him
attributed this to an intermittent nervousness which he never
managed completely to repress. There was the occasion of the
presentation of the prizes following a regimental race when
Haig said to the winners: 'I congratulate you. You have done
very well. I hope you will run as well in the presence of the
enemy.' There was the speech he made at Cambridge before
King George, high political and scholastic dignitaries and
officers. Haig, for some reason, ignored his written notes to
utter in strained tones strings of sentences which, to the audience,
had no sequence or meaning. As the speaker rambled on,
some of the older members of his audience fell asleep. Then,
when war came, smiles were repressed when one day Haig
uttered this startling and original thought: 'We cannot hope

to win until we have defeated the German army.' Trying
manfully in France to make conversation with a humble
private, Haig asked him: 'And where did you start this war?'
The embarrassed private replied: 'Nowhere, sir. I didn't start
no war.' However, Haig was impatient of woolly or vague
conversation from others. He interrupted a statement being
made to him by one young officer with, 'Don't ramble. Say
what you've got to say.' To another subordinate who was
describing a wartime occurrence with drama and emotion,
Haig snapped, 'Don't be a damned fool. Stick to the facts.'
Emotionally excited by battle, a staff officer addressed Haig in
unusually high-pitched tones. 'What's the use of an S.O. who
squeaks?' Haig remarked later. 'If he squeaks like that, what
will he do if there is real fighting?'

Wartime prime minister David Lloyd George sneered that
Haig was 'brilliant to the top of his boots'. Lloyd George
thought him 'a confused talker and never a clear thinker'.
What then could have been the reasons for the faith eventually
placed by much of the British nation in Haig, and for his
unchallengeable power and command over all ranks of the
army who held him in awed respect? Up to his command at
Aldershot and to the outbreak of the First World War he had
been an officer of top-rate efficiency. He almost re-created the
Indian army. His planning work with Lord Haldane years
before the First World War made possible the swift and
successful departure of a highly trained expeditionary force to
France at the start of the great struggle. He was magnificent at
training troops, at imbuing them with self-confidence and
pride. Haig could be a hard man, but he was invariably fair.
He created the impression of being a man of great integrity,
but in truth he could be tricky and a schemer when pursuing
his own ambitions. Of course, his close friendship with the
king consolidated his power and progress. Then Haig caused
men to respect and reverence him by a manner that was distant
and superior. The late H. L. Mencken described 'the unnatu-

rally superior air, a talent for sniffishness and reserve' as 'the most valuable asset any man can have in the world'. Added Mencken: 'The generality of men are always greatly impressed by it, and accept it freely as a proof of genuine merit. One need but disdain them to gain their respect.' However, when, with Aldershot in his past, Haig became commander of the British forces in the titanic struggle against Germany he was challenged by strategic problems which were beyond his intellectual capacity. These problems could not be sniffishly ignored or disdained. Thus Lloyd George's assessment of his ability may have been close to the truth.

At Aldershot Haig's obsession for the maintenance of his idea of a high pitch of efficiency through riding and other manly sports had its amusing side. For example, there was his grand assault upon the *thé dansants*. There was born, before the 1914 War, a craze among young people for making teatime an occasion for the dance. In the best hotels, restaurants and tearooms of the large department stores, the girls and young men gathered at tables around a floor cleared and brushed with french chalk for dancing. At four or thereabouts a small orchestra would start to play. Couples would rise and whirl around the floor in waltzes, polkas and foxtrots. At the end of each dance the young men and girls resumed their consumption of tea and fancy cakes. It was all apparently decent and harmless. Nevertheless *thé dansants* received lots of criticism from serious, older folk who saw them as first steps towards a maiden's downfall. The real worry of the critics was not the *thé dansant* itself but what a couple might be excited and stimulated to do *after* it. Moreover, at a time when there were many reports about vice and 'white slavery', there were rumours that the 'slavers' were haunting *thé dansants* with the object of luring girls away to brothels. Among other accusations made against the *thé dansants* was that many of the young males patronising them were dandies, milksops and generally symbols of moral decay. Indeed, it was said that hundreds of young men were forsaking their Saturday football, hockey or lacrosse games to mince and dance with flirtatious girls in an enclosed and unhealthy atmosphere.

The day came when subalterns of Aldershot's Officers' Club wanted *thé dansants* too. The committee thought this a good idea. Thus soon an orchestra was often playing in the afternoon in the club while officers and girls followed the ritual of tea and dancing that was sweeping through England. The air was fragrant with the aroma of Egyptian cigarettes, so fashionable at the time; also with the exotically scented sachets and cachous of the young females pursuing the ultimate in glamour. Sometimes while they danced the couples would sing the chorus of the contemporary pop tunes being played, such as:

'Mabel dear, listen here,
I'm afraid to go home in the dark.'

News of these affairs reached the G.O.C. What had been criticised as 'decadent goings-on' shocked Douglas Haig, and his reaction was dramatic and incisive. He commanded that these *thé dansants* cease forthwith. The duty of officers, he said, was to pursue recreations which were manly and soldierly, particularly during the hours of the afternoon. Instead of indulging themselves in dances with young women in a stuffy room they should be golfing, riding, playing tennis or vaulting on the parallel bars in the gymnasium. Of course Haig did not object to dancing at formal balls held at the proper time—the night—and with the accepted accompaniments of programmes and chaperones. Indeed, in India he had attended many balls and masquerades too, once in the costume of King Henry VIII. The happy and innocent hours of the *thé dansants* which Haig ended nevertheless survived in the memories of the subalterns and captains until blacked out with all other recollections when most of these young men perished so soon afterwards on the battlefields of France and Belgium.

Haig, while disdaining such fads as dancing at tea, would encourage any other which might, even to the point of eccentricity, require strength and endurance. Thus in April 1913 forty-year-old Tom Burrows, Australian club swinger, stood with the G.O.C.'s hearty approval in the Wellington Lines gymnasium continuously swinging his 3 lb. 6 oz. clubs for 107 hours. Burrows, in a much publicised performance which excited Aldershot civilians as well as the military, started swinging at six on a Monday night to the accompaniment of the bands of the King's Royal Rifle Corps and Hampshire Regiment. 'Three drummers', says a contemporary description of the big non-stop swing, 'agreed to remain with Burrows all night, while Lieutenant Green also beat the big drum. Bands and drums also played throughout Friday night and Saturday morning until the dramatic finish took place.' But actually there were, during the whole period, drummers stationed close to Burrows. They started banging hard whenever the swinger showed signs of fatigue by closing his eyelids. And once Professor Leonard Hill, of the London Hospital, gave the swinging Australian oxygen while the crowd of spectators chanted: 'Stick to it, Tom! Stick to it!' In the final hours of Burrow's performance mobs inside and outside the gymnasium yelled songs. Afterwards he was examined by doctors who announced that his health was not affected. Then he collapsed and slept continuously for twenty-eight hours.

Among the more serious activities of The Camp in 1912 and 1913 were the many lectures and discussions on the current Balkan Wars, their battles and their strategies. The wars seemed far away and also remote from the interests of Great Britain. But they had a picturesque Ruritanian quality, with four bewhiskered monarchs in *Chocolate Soldier* uniforms active in the field. These were Kings Ferdinand of Bulgaria, Peter of Servia, George of Greece and Nicholas of Montenegro. The Balkan Wars at a distance had a cinema-like quality, but they did not appear to equal in realism D. W. Griffith's film *The*

Birth of a Nation, the 1913 American Civil War epic that fascinated our own troops.

The possibility of a great war actively involving Great Britain engrossed the War Office and, of course, Douglas Haig; but the possibility was considered remote by the ordinary officers and privates of The Camp. The nation certainly had her troubles with the Irish question, suffragettes, Socialists, working-class poverty and the vexed controversy over the disestablishment of the Church in Wales. However, soldiers and citizens saw their country as all-powerful and impregnable. Certainly she had her slums and her poor; but nevertheless she was fantastically rich, with even the United States owing her money. Her private investments abroad were in the neighbourhood of £400,000,000, thus bringing in an enormous income. Britain had the biggest and most powerful navy ever built and also the biggest mercantile marine, despite the German kaiser's attempts at rivalry.

Undoubtedly the big majority of those who dwelt in The Camp and Aldershot town were more interested in local vogues and progress than in any gossip about martial threats from overseas. Diabolo was a tremendous local craze. Everywhere in the open air men, women and children were sending their reels spinning into the air from a string attached to two sticks. The new Aldershot Hippodrome became the most popular show place in town, with two variety shows nightly and advertising that 'prices from 3*d*. can be booked by phone or telegram'. The popularising of the 'free wheel' led to the proliferation of pedal cycles, despite the growing challenge from motor cycles and motor cars. Prices for bicycles in Aldershot in 1913 ranged from four pounds to sixteen guineas. Far quieter and stealthier than motor cars, the uses of pedal cycles in future wars were discussed by advanced military thinkers in The Camp who studied the exercises of cycle detachments of troops there with close interest. But troops on bicycles failed to excite the cavalry-minded Douglas Haig.

However, the 1913 edition of the popular but authoritative *Everyman Encyclopaedia** pointed out that 'Cyclists are cheaper to maintain than cavalry or mounted infantry, are more easily trained, can cover longer distances at greater speeds, and are silent in their operations. They are of great value in conveying messages, in seizing quickly upon unoccupied positions, in reinforcing weak columns. The introduction of the motor car for pleasure purposes is seriously affecting the reserve supply of horses upon which the various governments can fall back in time of emergency, and therefore greater importance will probably be attached in future to the proper organisation of cyclist corps.'

Aldershot started its 1914 summer without a hint of foreboding of what was to come. There was the usual period of rigorous annual training under the sharp and stern supervision of Haig. A very minor but human feature of this was a battle between the Officers' Training Corps of Oxford and Cambridge Universities near the Basingstoke Canal. A begrimed and sweating undergraduate in charge of the scouts of the Oxford corps was seen being bawled out by a huge sergeant-major. This undergraduate was Edward Prince of Wales, who was to become king and after that the Duke of Windsor.

The effectiveness of the 1914 summer training of the troops was, it was said, soon to be thoroughly tested in the usual manœuvres. Meanwhile Haig and others were studying a new mobilisation scheme without any belief that this was so close to being put into operation. Brigadier-General John Charteris in his memoirs† recalled too that at this time 'some important polo matches were pending'. On August 1, 1914, Germany declared war against Russia, and two days later against France. The German army invaded Belgium, and on August 4 Great Britain declared war. In the words of Foreign Secretary Sir

* Published by J. M. Dent & Sons, London.
† *At G.H.Q.*, published by Cassell, 1931.

Edward Grey 'the lamps were going out all over Europe'. But at the War Office and at Aldershot Command headquarters they burned all night as plans long made for the eventuality of war were prepared for action. On the following day Haig hurried to London for the Council of War called by Prime Minister Herbert Asquith. And Aldershot received its detailed mobilisation orders.

Faithful to their traditions, the townsfolk of Aldershot responded to the call of war with hysteria and many un-balanced actions of 'patriotic' emotionalism. But the display was greater and generally sillier than those that greeted the outbreak of the South African and other wars. The public houses not only rang with popular songs, but people threw themselves into wild dances to the tunes, according to the memories of local residents. Women and girls, who were usually quiet and self-controlled, grabbed strange soldiers in the streets and kissed them fervently. At the Theatre Royal, actor J. Darley, playing the role of the priest in *The Rosary*, stepped before the curtain in full clericals to bless troops in the audience and to wish them 'good luck and a speedy return from the war'. At the Hippodrome tears coursed down the cheeks of men and women as they stood and sang 'Rule Britannia' and 'God Save the King' to the music of the grand orchestra. On August 11 King George and Queen Mary spent the day at Aldershot. Over luncheon with the Haigs and a few others the monarch expressed pleasure that Sir John French would command the British Expeditionary Force on the Continent. Then the bosom friends, King George and Douglas Haig, sat close to each other on Government House's terrace overlooking its lawns and flower beds rich in colour. Over the coffee and brandy Haig softly expressed his doubts to his friend regarding the fitness of French for such a difficult and respons-ible post. The mischief-making had started. The seed had been sown for the evil feuds of the generals of the First World War.

Soon after midnight on August 13 a large body of troops left the barracks of The Camp on the first stage of their move to Belgium and France. Their departure was supposed to be secret. But all Aldershot had somehow learned in advance about it. Scenes at the start of the South African War were repeated as the men marched from barracks towards the railway station. Soldiers' wives and children from married quarters clung to them, and some had to be dragged forcibly away. Others ran crying behind the troops. Waiting at the station in the darkness were hundreds of Aldershot townsfolk to add to the fervour of farewells, with the monstrous prophecy that became general on such occasions, 'You'll be back home before Christmas!'

Next many trains were arriving at Aldershot filled with men who had arrived for preparation and training before proceeding overseas. Immediate accommodation was arranged for twenty-four battalions of infantry and other troops. On August 14 Haig had left to command the First Army Corps of French's Expeditionary Force and to steam to France with other high staff officers in the liner *Cambrie Castle*. His wife had provided him with a luncheon basket as if he were off for a day of shooting game or deer-stalking. Haig was succeeded as Aldershot G.O.C. by Major-General W. Archibald Hunter.

Meanwhile the hysteria in town and camp grew worse with an epidemic of spy mania. Soldiers and civilians expected to find a spy behind every tree, around every corner. When one respected twenty-eight-year-old Aldershot resident was on a quiet evening stroll past a bridge he failed to hear the challenge of a sentry. The sentry shot him dead, and was later distressed to discover that his victim was his own brother-in-law. The sentry, a member of the Territorial Army, had, it appears, become over-excited by warnings that the Aldershot neighbourhood was teeming with spies. On another day a regular soldier shot dead a comrade who happened to be passing a powder magazine in The Camp. 'I couldn't see him clearly,' explained this soldier at the inquest, 'but I felt certain at the time he that was a spy.'

Aldershot had long been the home of numbers of quiet, respectable shopkeepers and other tradesmen of German nationality. The zealous citizenry classed them all as spies and demanded their detention. Hundreds of circulars, authorship unknown, were handed to passers-by in the streets and dropped through letter-boxes. These read:

'Britishers! To support a German or an Austrian is an act perilous to your own interests and to patriotism. We have in this locality German retailers such as bakers, butchers, hairdressers, etc., who trade to the detriment of our kinsmen. Beware of the peril! God save the king!'

However, all the enemy aliens of Aldershot were already under secret police surveillance. The war was still young when constables and detectives, penetrating into the food shops and the hairdressing saloons of the frightened Teutons, marched them away to heavily guarded internment camps. Among these prisoners was a German known as 'Fritz' who cut the hair of many soldiers in The Camp itself. 'Fritz' was a popular character there until he was seized for internment. Then a story circulated that, in between haircuts, he would furtively draw maps on shaving papers to be smuggled to the German High Command in Berlin.

Within a short time of the outbreak of war most of the 'regulars' had left Aldershot. They were replaced by thousands of the new recruits—members of what was known as Kitchener's Army. They had responded to the fantastic-looking posters of Kitchener, Secretary for War, glaring over his big handlebar moustaches, pointing with his index finger and saying that the country needed them. Today this picture which had such an appeal to the innocent idealism of the youth of 1914 is a derisory souvenir offered for sale in the vulgar shops of London's Carnaby Street. The 1914 volunteers were a motley mob, with members of all classes, filling the streets of Aldershot with simple and appealing fun and laughter, as if they were choir boys on a day's outing to Brighton or Southend. Only a few went after the drink and the fleshpots of the town. Very elderly residents in Aldershot still remember them as sweet, gallant and high-hearted idealists with little resemblance to the troops who had come to The Camp in all its preceding years. These recruits wore civilian clothing, for no uniforms were ready for them until the early months of the following year.

Towards the end of September 1914, King George, his queen and their daughter, the Princess Mary, came to see the volunteers, accompanied by Lord Kitchener himself. They arrived with members of the court and others in a cavalcade of cars and stayed several days in the Royal Pavilion. The king spent much of his time looking around everywhere in The Camp without advance warning or ceremony. A man who was there at the time told how, when the king's face suddenly appeared around the open door of a hut, one of the new soldiers exclaimed in his amazement, 'Christ, it's the bloody king!' before recovering himself and snapping to attention. On more formal occasions during their stay the king and queen inspected the new 15th (Scottish) Division and the 14th new (Light) Division. The lines of Scots on parade stood at attention in only shirts and trousers, mostly held up by braces, and in cloth caps. Most of the men of the 14th Division wore woollen jerseys or pullovers. The majority covered their heads with cloth caps, but among them were new soldiers in lounge suits, stiff white collars and trilby hats.

The first war-wounded had been brought to the Cambridge and Connaught Hospitals at Aldershot towards the end of August 1914. The Empress Eugénie was preparing to turn her home at nearby Farnborough into a convalescent home for wounded officers. Everywhere hospitals were being expanded, and many additional ones were being jerry built with wooden materials. The war was not over and 'the boys' were not home

by Christmas as so many simple folk had expected. Nevertheless there was widespread belief among civilians that the
geniuses of France's General Joffre and Britain's General
French would, through brilliant strategy and manœuvre,
smash and overwhelm 'the Huns' and thus quickly end the
war. What actually happened still brings searing grief in
reflective moments to many of those who remember it. When
French's strategy failed, Haig, who had done reasonably well
as commander of the First Army, was appointed his successor.
He was appointed new commander-in-chief in 1915 after a
series of military and political intrigues, the details of which
make scandalous and almost obscene reading in the light of
today. Haig had written secret letters to his friend King
George regarding the shortcomings of French, his former
friend. And the influence of the king behind the scenes played
a big part in the removal of French and the substitution of
Haig.

Douglas Haig as commander in India and Aldershot made,
as we have seen, an admirable contribution in bringing his
troops to a high pitch of training. He with other British generals
created a magnificent human fighting machine. The tragedy
was that they did not possess the imagination or mental
flexibility to know how to use their fine and fearless troops
with decisive effect. From the start of the First World War
Haig, with defeatist zeal, always insisted that it must last for a
long time. He believed in the inevitability of 'long fluctuating
battles', of a process of 'wearing down' the enemy by the
relentless and merciless sacrifice of the lives of his men. 'Germany's resistance,' he preached, 'must be worn down by a
continuous battery on her frontier . . . until the time comes for
the delivery of the decisive blow.' He directed this 'battery'
from luxurious château or villa headquarters, in cosy isolation
from the scene of battle. Surviving diaries of guests who were
entertained there contain comments on its dinners that would
have made even the mouth of General Sir Redvers Buller

water. Then from his grand headquarters Haig made big strategic decisions, involving the lives of hundreds of thousands of soldiers, without consultation with the ordinary fighting officers and men who were at the scene of battle.

Haig's Battle of the Somme was the greatest of World War I. The Germans were forewarned of the offensive by a preliminary bombardment lasting a week. This bombardment rendered great assistance to the enemy by churning up the ground, destroying the ease of any advance. The Germans held all the commanding positions. Haig stubbornly insisted in carrying on this offensive when it became obvious that it could never succeed. It lasted from July 1 until November 13, 1916. The British casualties of 343,000 exceeded those of the Germans whom Haig was supposed to be 'wearing down'. More than 90,000 men and youths, the finest of British manhood, perished in the battle, herded to slaughter like cattle in the stockyards of Chicago.

Then there was Passchendaele, from July 31 to November 6, 1917, fought by Gough's Fifth British Army, planned and supervised by Haig well behind the front. Once again the element of surprise was destroyed and the terrain whirled into an ocean of deep mire by a preliminary bombardment lasting ten days. There were 250,000 British casualties at Passchendaele, this second sample of Haig's strategy of 'wearing down' the enemy.

Still, because the entry of the United States into the war enabled the Allies to have more men to be killed than could the Germans, the war ended with 'triumphal victory' in 1918, with Haig serving under the supreme command of wiser Generalissimo Marshal Foch. And the British nation, for reasons which today seem inexplicable to leading military students, welcomed Haig home as a hero, showering him with honours. He was created Earl Haig, Viscount Dawick and Baron Haig of Bemerysde, Order of Merit. The House of Commons, who later treated the demobilised tommies with despicable meanness, voted Haig a gratuity of £100,000 to add to his whisky fortune. From a public subscription he was presented with the ancestral home of the Haigs at Bemersyde.

Haig, it is said, was a sensitive man. Indeed, during the war he avoided visiting casualty stations behind the lines because they made him feel so ill. He devoted the rest of his life to the British Legion, and to the welfare of all soldiers, with special solicitude for those who were permanently maimed. He instituted Poppy Day. A recurring duty was to appear at the dedication of war memorials in cities, towns and hamlets throughout the United Kingdom. As the sensitive man, he must have again and again been cruelly reminded of his strategy of 'wearing down the enemy' as he saw the many names on these memorials. He died on January 30, 1928, sitting undressed on his bed.

And what of Smith-Dorrien, the popular predecessor of Haig at the Aldershot Command? As commander of the Second Army Corps he won an impressive tactical success at Le Cateau during the retreat from Mons, but soon afterwards he fell out with difficult General French and was sent home. From 1918 until 1925 he was a beloved governor of Gibraltar. This officer who had so often risked his life on horses died in 1930, after a motor accident on the Bath Road.

The Camp at Aldershot was sixty-four years old when the First World War ended. And it was realised that it could never be the same again as slowly the tragic lessons of that struggle were comprehended. 'The bemedalled commander, beloved of the throne', could no longer attract such devotion and respect from his troops as 'he rode cock-horse to parade'. For the high officers, bred to the hallowed Aldershot tradition, had proved themselves unworthy. They had in the war pursued outdated and reckless strategies without regard for the lives of those hundreds of thousands of men whom they sacrificed with such cruel and unnecessary prodigality. The troops, from private to the ordinary non-staff officer, could be fooled no longer. There were so many soldiers back from the battlefields who witnessed the useless carnage, and so many military experts, led by Captain Liddell Hart, convincingly to indict the performance of the generals. Real respect of the troops for their high officers did not return until the Second World War when such commanders as Montgomery, Wavell and Alexander led their armies with strategic and practical skill, at the same time doing all possible to keep loss of life among their men at a minimum.

10

Lo' All Our Pomp of Yesterday...

W HEN the First World War ended the saying that 'things will never be the same again' became a platitude. Yet in Aldershot the truth of this statement when applied to The Camp was pitifully obvious. The army was like a man who had undergone a long and terrible illness. It was still alive, but physically and mentally changed. The war had been won. Nevertheless the army had taken a terrible beating. A million soldiers of the British Empire were dead. The army staff—those dashing officers who in overwhelming majority were of good family, good school, good riding and good sport—had been found badly wanting. The surviving rank and file who had once accepted their leadership with doglike trust, affectionately and without questioning, were disillusioned. In modern warfare, the staff officers whom they had regarded as their betters, had shown themselves at their worst.

The international freemasonry of royalty, rank and class which on visits contributed so much to the prestige and grandeur of Aldershot, had been broken by the First World War. The British royal family's network of relatives, from the German kaiser and Russian czar to countless continental princelings and grand dukes, was now removed from the international scene, leaving only a few pale monarchs in Scandinavia, the Balkans and elsewhere. And except for the French and Americans, who were never really welcome, there were no high officers from abroad worth being fêted in The Camp with its magnificent parades. Then the war had taken the glittering paint off the troops, like tin soldiers ill-used in a nursery. The gorgeous full-dress uniforms and trappings of the cavalry had ceased to be general and the horse, as an essential feature of the army, was moving towards its doom. Then the resplendent coloured infantry uniforms which had brightened the Aldershot picture for so long were replaced by drab khaki.

War memorials, multiplying from 1919 onwards, often named the holy trinity for which men were said to have 'laid down their lives'. This British Three in One and One in Three were God, king and empire. There were, as declining church-going figures showed, doubts about God after the 1914–18 holocaust. There was also growing cynicism or questioning regarding the ethics of empire. Thus only one member of Great Britain's trinity seemed unscathed. This was her monarch George V. While other thrones tumbled like pins in a bowling alley, the little king, supported by his big Queen Mary, strove to keep his wavering but still loyal subjects in the traditions of the old trinity in its entirety. In the life of George an important feature of the old-established order was Aldershot. When the First World War ended the king made it clear that he wanted The Camp to continue its grand and spectacular traditions and observances. He was the last monarch to have periods of residence in the Royal Pavilion. Indeed, he may truthfully be regarded as the last of The Camp's Old Guard.

Aldershot had played a significant part in the life of King George since he was taken there as a child by his grandmother Queen Victoria. Frequently as Prince of Wales he had stayed at the Royal Pavilion and participated in parades and reviews. Of his many Aldershot memories was that of the big Coronation Review in June 1902. On the eve of that review his father Edward VII was suddenly taken ill in the Royal Pavilion and hurried back to London by ambulance. There was panicky talk about cancelling this review until George suggested that his mother Queen Alexandra should take the salute. And in the uniform of colonel-in-chief of the Royal Fusiliers he took a

proud part in the parade past the queen. Through the years as
Prince of Wales, and then as king, George became increasingly
a familiar figure in the Aldershot scene. And now, after the
First World War, he was determined to preserve The Camp as
the show place of the army.

The first of the Aldershot 'spectaculars' after the war was
staged at King George's initiative for the Shah of Persia in
November 1919. Persia was then a chaotic country, and it
seemed hardly worth while for The Camp to make a lot of
fuss over her roly-poly monarch Ahmad Shah. But Foreign
Secretary Lord Curzon, with special imperialist interest, was
intent upon reorganising the Middle East and Near East to the
British Empire's advantage. Therefore 1919 saw the signing of
the Anglo-Persian agreement. In this Britain recognised the
'independence and integrity' of Persia in exchange for British
military advisers, a loan and help in improving internal communications. Ahmad Shah was thereupon invited over to
England on a state visit when every endeavour was made to
impress him with the continued might of the empire despite
the blood and treasure expended between 1914 and 1918.

On Monday, November 3, 1919, the shah was taken by
Prince Albert (later King George VI) to Aldershot to be wel-
comed by its G.O.C., General Sir Archibald Murray, at
Government House, and then to be honoured by a 'march
past' that would, before the war, have been worthy of a
German kaiser or a Russian czar. The troops consisted of three
regiments of Hussars, three brigades of infantry and a mixed
brigade of artillery. The shah was lunched and toasted at the
Officers' Club, then whisked by Albert to Farnborough
Aerodrome and Sandhurst for more inspections and parades.
The shah was visibly impressed by the British troops, guns and
aeroplanes. Yet Curzon's old-fashioned imperialistic scheming
was two years later ruined by a daring Persian army officer
named Reza Khan. After a successful *coup d'état* Reza Khan
called in Teheran a new National Assembly. This declined to

ratify the Anglo-Persian agreement. And in 1925 Ahmad Shah, over whom Aldershot had made such a fuss, was ignominiously deposed.

For some time there was a dearth of distinguished foreign visitors to The Camp, despite King George's enthusiasm in retaining its prestige. Prince Aage of Denmark stayed there for a few days early in May 1921. Then preparations were hastened to welcome a genuine V.I.P.—Crown Prince Hirohito of Japan. However, before that important occasion the town of Aldershot suffered an assault from The Camp unprecedented in its history. In the past, troops *en masse* had never made war on civilian Aldershot, although in The Camp they had several times fought fiercely among themselves. For example, in 1856, in the infancy of The Camp, British troops had battled with hundreds of men of Germany's 1st Regiment of Jägers who were quartered close to them. Then in the 1890s The Camp was several times disturbed by troops of rival regiments squabbling among themselves. The biggest incident of the kind was on May Day 1893 when men of the 20th Hussars and other cavalry attacked the barracks of the Cameronians in reprisal for the beating up of one of their comrades. A Colonel Long of the Cameronians was injured in the attack. This resulted in severe punishment of many concerned in the affray, and in the transfer of the Hussars and Cameronians to quarters far apart. The town of Aldershot had been spared such scenes. All its townsfolk had to endure from the troops were brawls in public houses and the occasional smashing of shop windows by soldiers temporarily uninhibited by drink.

However, on the night of May 7, 1919, men of The Camp made the first organised attack on the townsfolk in the history of Aldershot. The attackers consisted of several hundred of nearly 9,000 reservists who had been called back into the army because of the national crisis caused by the great coal strike and by the gravity of the situation in rebellious Ireland. The majority of the reservists were men disillusioned by the war, during and after which their ideals of 'God, king and empire' had evaporated. A woolly-thinking proportion of these now saw the red flag as the last hope for humanity. Thus in The Camp some of the reservists, rebellious and resentful, plotted 'a rising' in the town. On the night of May 6, in a preliminary sortie, some had bruised Superintendent W. Davis of the Aldershot constabulary and knocked out two of his teeth. One attacker had boasted that 'soon the red flag will be flying over this town'.

On the following night the reservists suddenly assembled in Aldershot behind a private waving a large red flag. For more than an hour they ran wild in Union Street, Wellington Street, Gordon Road and Victoria Road. Being a Saturday, there were many townsfolk about at the time, coming from the cinemas, music halls and public houses which were just closing. They were chased by the 'rebels'. According to a contemporary report, the town 'rang with the hoarse cries of men and the screams of frightened women'. The invaders from The Camp attacked and captured a motor bus. A soldier got on the top, waved a red flag and gave a rallying cry of 'Come on the rebels!' Windows of more than sixty shops were smashed and their goods looted. Running riot in a jewellery store, men stuffed their pockets with jewellery, diamond rings, watches and silver goods. Seizing clocks they hurled them at fleeing townsfolk. Meanwhile Aldershot police telephoned The Camp for help. Cavalry and infantry picquets arrived to make their way through streets deep in debris and broken glass. The rioters fled. Numbers were later identified by their bleeding arms and fingers caused by the window smashing. There was a long inquiry into this night of terror, and the Aldershot Command vowed that such a situation would never again be allowed to arise. However, there was a repetition on a smaller scale twenty-four years later, on July 4 and 5, 1945, when Canadian troops ran riot in the town, smashing the shops, but leaving civilians unmolested.

With many troops in a sulky and uncertain mood, the visit
of the Crown Prince, planned for May 17, 1921, was anticipated in The Camp with some trepidation. For special efforts
were being made to impress Hirohito wherever he moved in
England on this, his first journey outside Japan. The war had
made Japan, as Britain's ally, rich and important. Her profits
from aiding the allies with arms and other goods, were estimated at nearly £400,000,000 and her gold holdings had
climbed to £300,000,000. Although the government was
ending the Anglo-Japanese Naval Treaty to appease America,
it was most anxious not to incur Japan's disfavour. Crown
Prince Hirohito arrived at Aldershot in the uniform of a
British general. The troops behaved well and he inspected
them with apparent interest. Yet the young man had practically nothing to comment about them or anything else. All he
did was to bow and to nod. Most of the talking was done by
two members of his suite named Prince Kan-in and Admiral
Takeshita. Entertainment offered Hirohito included a gym-
nastic show and a sabre versus bayonet combat between
Quartermaster-Sergeant James and Sergeant-Major Reid.
Hirohito watched this clashing of sabre and bayonet through
his thick spectacles with oriental tranquillity. Next the G.O.C.,
General the Earl of Cavan, escorted the visitor and his entourage to The Camp's bakery to watch a loaf being made 'from
start to end'. On being told that the bakery turned out 60,000
loaves of bread daily for the troops, Hirohito at last spoke.
'Then your soldiers have good appetites?' he asked, only to
be assured by the earl that yes, the appetites of the troops were
good. A luncheon was given for Hirohito at the Officers' Club
where again conversation was sparse. When, at the conclusion
of further fatuous and stilted proceedings, the Crown Prince
left Aldershot, his hosts agreed that unfortunately this could
hardly be described as a memorable day. Thus there is good
reason to doubt whether today the Emperor Hirohito ever
recalls May 17, 1921, as he sits in his moated Tokyo palace.

Now let us look closer at King George V as last of the Aldershot Old Guard. After the Second World War he and Queen Mary made an annual stay at the Royal Pavilion as well as often visiting The Camp for the day on special occasions. The king made such an impression that people—now elderly—who were there at the time in the 1920s and 1930s have vivid recollections of his personality and of little incidents in which he was involved. Strange, but the king at Aldershot seems to have differed from the king known to the statesmen and society of London. Certainly he was often described as 'the sailor king' because of his youthful years in the Royal Navy. Nevertheless at Aldershot he was very much the old-fashioned army officer, fussy over ceremonial and tetchy when things went wrong. At times, indeed, his vocabulary of expletives could be as colourful and explosive as those of George Duke of Cambridge or Lieutenant-General the Honourable James Yorke Scarlett. At the same time he possessed the kind of rock-like faith in God as did the Victorian and Edwardian generals who had dominated the Aldershot Camp. This was the God worshipped in church parades as the special friend and benefactor of British arms and empire; as the Jehovah who would ensure their lasting predominance in the world order. Despite the awful British losses of the war, George's faith in his Jehovah's leadership and beneficence never wavered. His faith was that expressed in his favourite hymn in which he would join with gusto at services in the Garrison Church of All Saints and elsewhere:

'Lead us Heavenly Father, lead us
O'er the World's tempestuous sea;
Guard us, guide us, keep us, feed us,
For we have no help but Thee . . .'

The king preferred sermons to be happy exhortations, devoid of discussion on complicated points of doctrine and dogma. He accepted the Apostles' Creed and Thirty Nine Articles of the

Church, of which he was head, without question. To him the 'truths' were so obvious that any intricate explanation about them from the pulpit was superfluous. Hence he could spare little time for any sermon. The troops had, of course, to be very smart and well behaved at church parades honoured by the presence of the king. But they were happily aware that his presence was a guarantee that the sermon would be short. For chaplains were warned in advance that his majesty wished the preaching to be limited strictly to ten minutes.

King George V was no intellectual in the modern sense of that word. This must have been partly why he felt so much at home in The Camp, associating with the remnants of an officer class who had attained their positions in the army through wealth and social distinction rather than through intellectual strength. In the post-war era, from 1919 to the death of the king in 1936, all the Aldershot G.O.C.s were 'polished gentlemen' of aristocratic background, from General Lord Rawlinson and General the Earl of Cavan to General the Honourable Sir Francis Gathorne-Hardy. In 1919 many war volunteers of mixed background who had won wartime commissions remained in the army as 'regulars'. However, the time had not yet come when clever lads from grammar, technical and secondary schools would be established in great numbers at the Royal Military College, Sandhurst, so long the exclusive preserve of public school boys, the sons of professional men and of those described as 'private gentlemen'. Despite the impact of the recent war, the officers of Aldershot were still preponderantly unsophisticated gentlemen, sharing King George's simple faith and ideals, together with his simple sense of fun.

George loved the type of joke today dismissed by the American word of 'corny'. Newspaper reports of the 1920s and 1930s testify how many times the king would 'laugh heartily' at such humour. Indeed, in those years hardly a week passed without a newspaper story of the king 'laughing heartily' at

P

this or that. For example, George was thrown into one of those fits of laughter when, at a Royal Command variety performance, a comedian asked, 'How do we know the king loves plums?', then adding: 'Because the National Anthem says "Send him victorias, happy and glorious".' This was the kind of fun common and easily comprehensible to the unsophisticated minds of officers and men of The Camp. No wonder that in or outside the Royal Pavilion the king had many opportunities for laughing heartily.

But there were those times—still remembered with awe by men who were there—when the king was in an imperious and difficult mood. This was when he saw himself, not as the folksy father of his people, but as imperial sovereign to whom all knees should bow. Queen Mary, as his loyal helpmeet, could also be relied upon to ensure from time to time that the royal family of the greatest empire in history be handled with precise respect. The king who so often laughed looked cold and stern one morning a few minutes after troops had paraded past him. These included Lieutenant-Colonel H. C. R. Green (later Brigadier-General Green), who led his unit with sword upraised in salute. After the 'march past' Green had replaced his sword in the scabbard attached to the saddle of his charger when he was told that his majesty wished to speak to him immediately. Green rode into the royal presence, dismounted without his sword and stood smartly at attention. The king glared at him for a few moments and then loudly inquired, 'Since when has it been your custom to appear before your sovereign without your sword?' The king turned away without another word. Blushing with shame, Green saluted, remounted his horse and rode away never to know the reason why George had sent for him. On another Aldershot occasion a mounted officer of the Royal Engineers was galloping on to the parade ground for exercises before the king when his hat blew off. Because of the movement of the troops he was unable to halt and retrieve it. He was quickly spotted by the king among hundreds of officers and men. 'What!' roared George. 'An officer on parade without a hat?' The offending officer was ordered out of range of the royal vision. Later he was severely reprimanded for permitting his hat to be blown off. As a punishment he was made to wear it with the chin-strap down for several months, much to the merriment of the company he commanded.

In the grounds of the training school for army nurses, which has replaced the demolished Royal Pavilion, there works an elderly gardener who remembers George V as one who was invariably kind to his domestic staff but often fierce with members of his family, Edward Prince of Wales (now the Duke of Windsor) in particular. The gardener would watch the king as he mounted his charger before riding out with his staff to the parade ground. The king on such occasions was a demon for punctuality. To ensure that he would neither be too early nor too late in making the grand appearance he had a small clock of great accuracy concealed in the saddle. The gardener recalls awful occasions when the Prince of Wales, due to ride out to the troops with his father, was late. The father would stare at his clock as the second of departure drew near, using strong language and vowing, 'If he isn't here within a few moments then he must be left behind.' And when the Prince of Wales, rarely a punctual person, did arrive the king, scarlet in complexion and language, would scold him.

Some officers who in the First World War had been calm and fearless in the face of a powerful and ruthless enemy were in Aldershot thrown into fear and confusion in their first social encounters with Queen Mary. The queen, when she so desired, would be charming; but too often she was inhibited by the strict observance of the ancient tradition that the subject must never speak to royalty until spoken to. This discouraged spontaneous conversation. It resulted in long silences between the queen's question or remark with the officer's reply and the queen's next question. The wives of officers were in Royal

Pavilion tea parties received graciously by Queen Mary. But all such gatherings had periods when everything was reduced to mime, as if a silent film were being enacted. Many was the time at Aldershot and elsewhere when the queen was at the side of the king while the troops marched by. As tall and upright as any guardsman, Mary could be critical of the performance of the soldiers they watched. After a 'march past' in Hyde Park she was heard to tell her husband, 'I've never seen the Guards march so badly. You'll have to speak to them about it.'

On June 21, 1922, the town of Aldershot became more than an adjunct of The Camp. The place that had evolved from the 1854 community of squatters became a borough. The Charter was unique in that it specified that of the borough's twenty-one councillors, three should be army officers appointed by the Minister for War. Actually the army had, since 1857, three representatives on the governing body of the town. And the army, parading in great numbers, seemed to steal the show in the 1922 Charter celebrations in spite of the new regalia in which the mayor, Mr. A. H. Smith, and his fellow aldermen had adorned themselves. In the Whitsun of the following year the king invited all the members of the borough council to the Royal Pavilion. He stressed his affection for Aldershot and the pleasure which residence in the Pavilion gave to himself and to the queen.

It was also during King George's visit of Whitsun 1923 that for the first time in an Aldershot royal review the technicians attracted bigger attention than the horsemen. Of course there had been tanks since 1916, so it was inevitable and almost essential that some specimens should roll past the king and queen at this, the first royal review in The Camp since the war. With them for the occasion George and Mary brought their son Prince Henry (later the Duke of Gloucester), their daughter Princess Mary (later the Princess Royal) and son-in-law Viscount Lascelles, Earl of Harewood. The princess had a year previously married Lascelles, who, although a former Guards officer and an upright character, was unpopular with the rank-and-file of the army. According to barrack-room gossip Mary had reluctantly been forced 'to marry Lascelles for his money'. This was absurd and untrue, since the king himself was so rich that there could never have been any temptation to sell his daughter. Then the inevitable jokes about the couple circulated in the barrack rooms and bars such as: 'Lascelles is a very rude man. He pinched the queen's little Mary.'

There were but eight tanks and eight armoured cars in this royal review of 8,300 troops, but they attracted the greatest interest and won special cheers from the crowds of civilian spectators. A few days later the king, mounted on his dark brown charger, watched tanks in action in training operations. They emerged through smoke screens put up by howitzers across the Long Valley of Aldershot. As Sir Basil Liddell Hart and other military experts have recorded, there were still at the time many high officers who regarded tanks as ineffective interlopers in the science of war. And their belief never wavered until Hitler's victorious blitzkrieg in 1940. Nevertheless, throughout the 1920s and 1930s tanks and armoured cars relentlessly reduced the prestige of the once glorified war horse.

After the First World War the British government had callously insisted upon the abandonment in the Middle East of 22,000 British army war horses on the grounds that shipping was short and that it 'wasn't economic' to transport them back home. Most of these were sold to natives into conditions of cruelty and hardship. Some 4,000 of these animals passed into the hands of Egyptian fellaheen, many being starved and worked to death by their ignorant new masters. All would have been doomed but for the intervention in the early 1930s of Mrs. Dorothy Brooke, wife of Major-General Geoffrey Brooke, then commander of the British Cavalry Division in Egypt. Shocked by seeing so many skinny and sore-covered horses bearing British army brands slaving in the streets of Cairo, Mrs. Brooke started a fund for buying them back. Incidentally, this led to the foundation of the Brooke Hospital for Horses in Cairo which today freely treats the horses and donkeys of that city's street traders.

In Aldershot and other army centres in the 1920s a large number of horses were being 'retired' rather than abandoned as the war machines replaced the war horses. In contrast to the atrocious fate of horses deserted in the Middle East, those in England were demobbed with affection and care, every effort being made to find them new homes where they would be well cared for. Heart-breaking to many cavalrymen, embued with faith and affection for the horse, were the final mounted parades prior to the mechanisation decreed by the War Office. In Aldershot on April 10, 1928, there was the last mounted parade of the 11th Hussars, the renowned 'Cherry Pickers' who won such lustre in the Peninsular and other campaigns. The 11th Hussars was, together with the 12th Lancers, to be reorganised and equipped as an armoured regiment. With Lieutenant-Colonel F. H. Sutton commanding, 250 hussars of all ranks paraded, but in drab khaki instead of the colourful dress uniforms of the historic and equine past. The parade was pervaded with sadness, some of the men struggling to suppress chokes of emotion. It was announced that officers would be allowed to retain their chargers 'temporarily'; also that a number of the horses would be transferred to the Royal Mews, the Duke of York (later King George VI) being colonel-in-chief of the 'Cherry Pickers'. After the parade a trooper broke the gloom with the remark: 'We got a name for picking cherries. I suppose that now, with our bleeding tanks, we'll get a name for picking up nuts.'

Aldershot's Age of Splendour had its grand and unforgettable finale on Saturday, July 13, 1935, in the Royal Review in celebration of King George's Silver Jubilee. This was a mighty royal occasion, never again repeated on such a scale. It would have delighted the Duke of Cambridge and all other martial figures of the Victorian and Edwardian past to whom spectacular marching, stirring band music and the gorgeous, precise rituals of the peacetime parade ground meant so much. Such 'spectaculars' were in future only to be partially imitated for the entertainment of the masses in the Aldershot Tattoos and other army shows. This Royal Silver Jubilee Review was, indeed, held in the great Rushmoor Arena, opened in 1922 to accommodate Tattoos and audiences numbering many thousands. And after that tremendous day of the review the grandeur of Aldershot began to fade. The approach of the Second World War and the war itself changed it into a more serious place, with fewer and fewer showy parade-ground diversions and with a decline in important visitors. Then in recent years The Camp has lost its solid if uncomfortable Victorian barracks for new erections, allegedly rich in utility but poor in grace and charm. Aldershot has also had its status reduced from the prestige of a Command to merely that of the South-East District of the Southern Command. It seems that everything possible has been done to destroy The Camp's traditional character since that summer day in 1935 when George V and the thousands around him witnessed its final blaze of glory.

The king and queen were not staying in the Royal Pavilion at the time of the Silver Jubilee Review. The Pavilion was being turned over to their son and daughter-in-law, the Duke and Duchess of Gloucester, while he was taking a course at the Staff College in nearby Camberley. And after that royalty was never again to occupy the quaint wooden house which the Prince Consort had designed with such enthusiasm after the foundation of The Camp. The king and queen drove to Aldershot from Buckingham Palace for the review. But there was surprise among the 50,000 spectators when Queen Mary appeared in the pavilion of the Rushmoor Arena without either the king or their sons. She was greeted by Minister for

War Viscount Halifax, who, like the late kaiser, had a
withered arm. Soon the queen was conversing with an old
man whom few in the crowd could recognise. This was
Arthur Duke of Connaught, aged eighty-five, a shrunken and
almost mummified remnant of Victorian and Edwardian
Aldershot. As General Officer Commanding at The Camp,
old Arthur had participated in so many grand parades and
reviews when Europe had so many proud monarchs to
impress and entertain. This favourite son of Queen Victoria,
who was always 'so good', was unwell. Nevertheless, he had
insisted on coming to Aldershot to watch the pageant of the
marching troops and to strain his rather deaf ears to hear the
music of the massed bands.

Eyes were moved from the queen and the old man towards
a mass of trees on the other side of the arena. There emerged,
on a chestnut charger, King George, erect and solemn, in the
khaki uniform of a field marshal. And then, behind their
monarch and father, there rode the Prince of Wales, the Duke
of York, the Duke of Gloucester and the Duke of Kent, each
in the uniform of his respective regiment. All were solemn like
the king, except the Duke of Kent, who was smiling. George
and his four sons were followed by a mass of army 'brass',
British and Indian, the highest ranking officer being the Chief
of the Imperial General Staff, Field Marshal Sir Archibald
Montgomery-Massingberd. Approaching the Rushmoor Pavilion, King George and his sons saluted the queen. The king
then dismounted and stood at the saluting base for the 'march
past' led by the Aldershot G.O.C. General the Hon. Sir Francis
Gathorne-Hardy. Later the parade halted before the sovereign
and lowered the colours to the tune of 'God Save the King',
played by the massed bands. Gathorne-Hardy called for three
cheers for the king, who within a few months would be dead.
For the second time the colours were lowered and the National
Anthem played. Aldershot's Age of Splendour had ended.
Now all its pomp belonged to yesterday.

Bibliography

Richard Aldington: *Wellington* (William Heinemann, 1946).

E. M. Almedingen: *The Romanovs* (Bodley Head, 1966).

A. Augustin-Thierry: *Le Prince Impérial* (Grasset, 1935).

R. N. W. Blake (Ed.): *The Private Papers of Douglas Haig 1914–1919* (Eyre & Spottiswoode, 1952).

Hector Bolitho: *The Reign of Queen Victoria* (Collins, 1949).

Victor Bonham-Carter: *Soldier True* (Frederick Muller, 1963).

G. A. Broomfield: *Pioneer of the Air* (Gale & Polden, 1952).

John Charteris: *Field Marshal Earl Haig* (Cassell, 1921); *At G.H.Q.* (Cassell, 1931).

A. Clark: *The Donkeys* (Hutchinson, 1961).

Howard N. Cole: *The Story of Aldershot* (Gale & Polden, 1951).

Georgina Daniell: *Aldershot: A Record of Mrs. Daniell's Work Amongst Soldiers* (Hodder & Stoughton, 1879).

A. Filon: *Le Prince Impérial, 1856–1879* (Hachette, 1912).

Roger Fulford: *The Prince Consort* (Macmillan, 1949).

Christopher Gibbert: *The Court at Windsor* (Longmans Green, 1964).

Alexander Godley: *Life of an Irish Soldier* (John Murray, 1939).

John Laffin: *Tommy Atkins* (Cassell, 1966).

Arthur Gould Lee: *The Flying Cathedral* (Methuen, 1965).

B. H. Liddell Hart: *Memoirs of Captain Liddell Hart* (Cassell, 1965, two vols.).

Edward McCourt: *Remember Butler* (Routledge & Kegan Paul, 1967).

Philip Magnus: *King Edward the Seventh* (John Murray, 1964).

C. H. Melville: *The Life of General Sir Redvers Buller* (Arnold, 1923).

Simon Nowell-Smith (Ed.): *Edwardian England 1901–1914* (Oxford University Press, 1964).

H. C. B. Rogers: *Mounted Troops of the British Army 1066–1945* (Seeley Service, 1959).

Giles St. Aubyn: *The Royal George* (Constable, 1963).

William Sheldrake: *Guide to Aldershot* (Sheldrake Press, 1859).

Horace Smith-Dorrien: *Memories of 48 Years Service* (John Murray, 1925).

John Smyth: *Sandhurst* (Weidenfeld & Nicholson, 1961).

Jane T. Stoddart: *The Life of the Empress Eugénie* (Hodder & Stoughton, 1906).

Julian Symons: *Buller's Campaign* (Cresset Press, 1963); *England's Pride: The Story of the Gordon Relief Expedition* (Hamish Hamilton, 1965).

John Terraine: *Douglas Haig: The Educated Soldier* (Hutchinson, 1963).

E. E. P. Tisdall: *The Prince Imperial* (Jarrolds, 1959).

Lawrence Wilson: *The Incredible Kaiser* (Robert Hale, 1963).

Evelyn Wood: *From Midshipman to Field Marshal* (Methuen, 1906).

Mrs. Young: *Aldershot & All About It* (Routledge & Kegan Paul, 1858).

G. M. Young (Ed.): *Early Victorian England* (Oxford University Press, 1951).

Anonymous: *Sketches of The Camp at Aldershot* (Andrews & May, 1858).

Index